THE WORLD

世界建筑事务所精粹

ARCHITECTURAL

FIRM SELECTION

OFFICE BUILDING / COMMERCIAL BUILDING /
CULTURAL BUILDING / COMPLEX BUILDING
办公建筑 / 商业建筑 / 文化建筑 / 综合建筑

深圳市博远空间文化发展有限公司 编

天津大学出版社
TIANJIN UNIVERSITY PRESS

PREFACE

PREFACE

In the past few decades, the practice of architectural design has developed towards the postmodern design concept with collage, mixture and coexistence from the modernism on the basis of functionalism and single typology. Rationality turns to be the theme of the development clue of architecture from the modernism stage dominated by the geometric form to the postmodernism stage with multiple mixing symbols. Both the architecture and the social activities of human penetrate and promote each other. The history, economics, philosophy, sociology, etc. have exerted unprecedented influence on the theory and practice of architecture design, which enable the contemporary buildings to obtain the ecological characteristics with multi-elements. The thoughts of architecture such as the phenomenology, semiology, structuralism and deconstruction endow the architecture in this era with symbolic interpretations and create the works for this era.

The innovative and pioneering architects open up new possibilities of the functions, space and forms of the architecture with nontraditional innovative spirit and creative practices. The design methodologies based on the information technology, such as the nonlinear design, parametric design and virtual construction, greatly enrich the design measures of the architects, and open the gate towards the architecture in the new era. The unrealized design creations such as non-standardization, structural skin, free form and surreal characteristics under traditional design conditions have displayed themselves with the development of the digital technology, information technology, structure and material technology. The architecture breaks through the restraints from the structure and function, and the shape and space tend to be more complicated and intriguing, displaying the unique characteristics and times imprinting of the contemporary architecture. Extensive excellent cases and architectural practices are provided in this book, which offers architectural cases with different functions, scales, fields and cultural backgrounds. Readers can spy into the influences and expressions of the above the mentioned architectural thoughts and design methodology in the excellent architectural creations, thus achieving thought-provoking inspiration and apperception.

Wang He
January 28, 2013

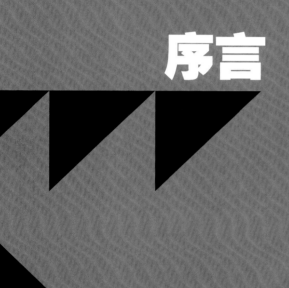

序言

在过去的几十年，建筑设计的实践从功能主义和单一类型学基础上的现代主义向后现代的拼贴、混合、共存并置的设计思想发展。建筑从强调几何形体构成的现代主义时期，到多元符号混合的后现代主义时期，理性成为这一发展线索的主题。建筑与人类的社会活动相互渗透、相互促进。历史学、经济学、哲学、社会学等都对建筑设计理论与实践产生了空前的影响，使当代建筑呈现出多元并存的生态特征；现象学、符号学、结构主义、解构主义等建筑思潮都给这个时代的建筑赋予了标志性的注释，并创造出属于这个时代的作品。

勇于创新和开拓的建筑师以颠覆传统的创新精神和创作实践，开拓了建筑功能、空间、形式的新可能。非线性设计、参数化设计、虚拟建造等后工业时代以信息技术为基础的设计方法论极大地丰富了建筑师的设计手段，打开了通向新时代建筑的大门。非标准化、结构性表皮、自由形体、超现实的特征等在传统设计条件下难以实现的设计创作，随着数字技术、信息技术、结构与材料技术的发展，一一呈现在世人面前，建筑自身突破了结构与功能的约束，造型与空间更趋于复杂与耐人寻味，表现出当代建筑独有的特征和时代印记。

本书提供了大量的优秀范例和建筑实践，呈现了不同功能、不同规模、不同地域和文化背景的建筑案例。读者可以从中窥探到以上建筑思潮和设计方法论在优秀建筑创作中的影响与展现，并从中得到思考性的启发与感悟。

王禾

2013 年 1 月 28 日

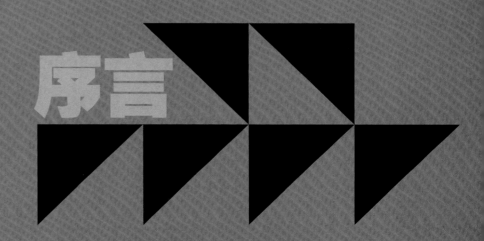
序言

目 录
CONTENTS

办公建筑　OFFICE BUILDING

Office building nowadays occupies a pretty high amount only second to residential house. Since we have entered a wholly new era of information due to the endless stream of emerging high–tech in the 21st century. The development of office building, as one of the main places to collect and handle information, is beyond comparison. According to an estimate, over a half of our population will work in office buildings in the middle of this century. So from this point of view, 21st century is "the century of office building".

As a kind of place to collect, handle and generate various administrative, scientific and commercial information, office building belongs to the infrastructure construction for social reproduction. It is a kind of place where information is produced, as well a place where we live. So the comfort level of the office directly affects the working efficiency of the staff. That is the reason why the requirement of a humanized design of a modern office building is exalted. Hence it is demanded that the design of the inner space of an office building should be as favorable to staff's working mind and behavior as possible. Similarly, a balance between the elegant, cozy and humanized outer and inner spatial environment is required. All these will greatly enhance the physical and psychological development of the staff, as well as the working efficiency.

Meanwhile, the spatial construction of office building is varying to adapt new official functions and new requirements for a higher comfort level. It is not only a request for a reasonable schedule of traffic routes, but also a result of spatial organization with a general consideration of elements such as aesthetics and technique. A mature and independent development system of office building construction is now gradually taking shape. Architects of office building are continuously seeking for brand–new design methods and architectural styles to capture the spirit of the era, and to create office works with the most era features and modern science and techniques.

It is gradually an important premise of architectural design to develop sustainably, environment– friendly and energy–savingly. How to maximize the function without the negligence of saving energy, and to make it a goal to be environment–protecting are among the key considerations of the architects. At present, the status of office building in the city is getting more and more important as its quality and quantity are increasing. It has become a significant part of the skyline of the city. Some momentous office buildings are even considered as landmarks of the city.

　　目前办公建筑已成为除住宅外，数量最多的一类建筑。在高新技术层出不穷的 21 世纪，人们进入了崭新的信息时代，办公建筑作为收集和处理信息的主要场所之一，其在信息时代的发展与过去已不可同日而语。据预测，到 21 世纪中叶，将有一半以上的人口在办公建筑里工作。从某种角度来讲，21 世纪可谓是"办公楼的世纪"。

　　作为收集、处理和产生各种行政、科研、商务信息的场所，办公建筑是社会再生产的基础性建筑。办公室既是信息生产的场所，又是人们生活的场地，其环境的舒适度直接影响着员工的办公效率，因此，当代办公建筑设计对办公场所的人性化要求也逐渐提高，这就要求办公建筑内部空间的环境向着最有利于人员办公心理和行为的方向发展，同时要兼顾内外空间环境的美观、舒适，充满人性关怀，这些对员工身心健康和工作效率的提高都能起到重要的促进作用。

　　同时，办公建筑的空间构成也在不断适应新的办公功能和舒适性要求的变化，这不仅是合理安排交通路线的要求，也是综合考虑审美、技术等多种因素后的空间组合的结果。办公建筑已逐渐形成了成熟而独立的发展体系，建筑师在不断寻求新的设计方法、建筑形式，以捕捉时代精神，创造出最具有时代特色，能够显示现代科技的办公建筑作品。

　　可持续发展和绿色环保理念逐渐成为建筑设计的重要前提，如何在实现功能最大化的同时实现节约能源、绿色环保的目标，成为建筑师在设计办公建筑时考虑的重点之一。如今，随着办公建筑数量的增加和品质的提升，它在城市中的地位也愈加重要，成为城市天际线的重要组成部分，尤其是一些重要的办公建筑，已作为城市地标而成为城市的象征。

FUKOKU TOWER
富国大厦

Architects: Dominique Perrault Architecture
Associate Architects: Shimizu Corporation Architects & Engineers
Engineering Architects: Shimizu Corporation Architects & Engineers
Client: Fukoku Mutual Life Insurance Company, Japan
Location: Osaka, Japan
Site Area: 3,900 m²
Building Area: 68,500 m² (including car park)
Photographer: DPA/ADAGP

设计机构：多米尼克·佩罗建筑师事务所
合作机构：Shimizu Corporation Architects & Engineers
工程设计：Shimizu Corporation Architects & Engineers
客户：日本富国生命保险株式会社
项目地址：日本大阪
占地面积：3 900 平方米
建筑面积：68 500 平方米（包括停车场）
摄影：DPA/ADAGP

This tower project for the Fukoku Mutual Life Insurance Company takes inspiration from the profile of a gigantic tree whose roots proliferate on the surface of the ground. Splayed at its base, the tower's outline tapers elegantly as it rises, gracing the city's skyline with a vertical asymptote. The contrast between the structure's base and upper regions is accentuated by the treatment of the building's "bark". Broad "wood chips" on the lower levels gradually give way to a sleek wall. The glass facade is worked into a crescendo of encrusted mirrors at the base, reflecting the colors of the sky and the surrounding environment. Situated at the exit of Osaka's main train station, the building will become a lasting landmark in the urban landscape.

　　日本富国生命保险株式会社的这个建筑，设计灵感来源于根系在地表扩散的大树的外形。大厦在基座上延展，以优雅的锥形轮廓向上延伸，以垂直渐近线的形态来美化城市的天际线。建筑立面像树皮，使建筑基座和上部区域的对比更加强烈。低层木片状的立面逐渐被光滑的玻璃幕墙所取代。玻璃立面像一面镶嵌在基座上的镜子，映射着天空和周围景观的颜色。大厦坐落于大阪的主要火车站的出口处，将成为当地城市景观的永久地标。

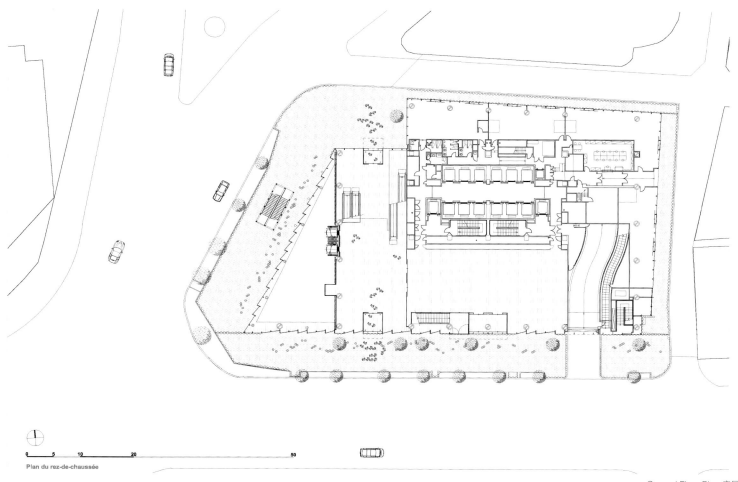

0 5 10   20   50

Plan du rez-de-chaussée

Ground Floor Plan 底层平面图

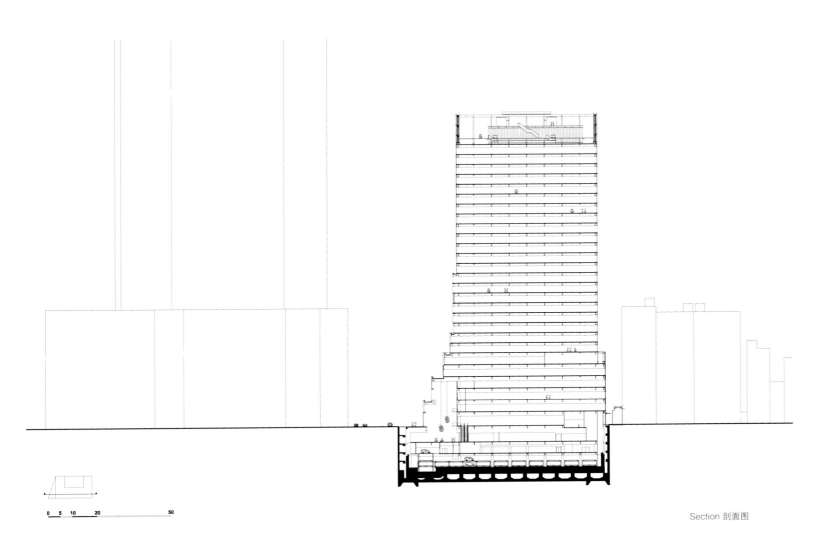

Section 剖面图

0 5 10 20   50

ILLA DEL MAR
ILLA DEL MAR 大厦

Architects: Adolf Martínez, Josep Lluís Sisternas(MSAAGROUP)
Jorge Muñoz , Enrique Albin(M+A Architecture & Planning Inc.)
Client: Espais Diagonal Mar
Project Manager: José Jurado
Project Team: Cristóbal Arrabal, Antoni Ramis, Tomás de Riba,
Ricard Morató(MSAAGROUP)
Location: Barcelona, Spain
Building Area: 60,377 m²
Height: 77 m. North tower, 89 m. South tower
Photographer: Adrià Goula, Rafael Vargas

设计机构：Adolf Martínez, Josep Lluís Sisternas
(MSAAGROUP)
Jorge Muñoz, Enrique Albin
(M+A Architecture & Planning Inc.)
客户：Espais Diagonal Mar
项目经理：José Jurado
项目团队：Cristóbal Arrabal,
AntoniRamis, Tomás de Riba,
Ricard Morató(MSAAGROUP)
项目地址：西班牙巴塞罗那
建筑面积：60 377 平方米
建筑高度：北塔 77 米，
南塔 89 米
摄影：Adrià Goula,
Rafael Vargas

Structural Architects: Agustí Obiol(BOMA)
Juan Pablo Rodríguez / Ignacio Vallet(MSAAGROUP)
Services Engineers: PGI
Yolanda Boto(MSAAGROUP)
General Contractor: Espais Promocions Immobiliàries (EPI)

结构设计：Agustí Obiol(BOMA)
Juan Pablo Rodríguez / Ignacio Vallet (MSAAGROUP)
服务工程：PGI
Yolanda Boto (MSAAGROUP)
总承包公司：Espais Promocions Immobiliàries (EPI)

The Illa del Mar building is part of an urban setting for Barcelona. The building is isolated and freestanding, situated on the coastal side of the Diagonal Mar city park. The nearest landmarks are the park itself, the main street that delimits it, and the rest of the buildings around the park.

Diagonal Mar is a distinctive project that seems to stand apart from the urban grid it interrupts. Open space and the park's orientation towards the sea are the parameters on which the ideas applied to the project were based.

Initially, in accordance with the urban plan for this area, the Illa del Mar was conceived as a set of three tall buildings. The most visible building would have more or less occupied the centre part of the edge of the park facing the sea.

Our proposal, adopted by Barcelona City Council, was to concentrate the building space in just two towers: the first, the lower one, aligned in continuity with other already existing buildings, situated on a line perpendicular to Avinguda Diagonal; the second, the taller of the two, on the sea front, displaced towards the end of the park, to allow an unobstructed and extensive view towards the sea from the park. The two towers are linked by a linear, low-rise structure, which gives the development the necessary unity without interrupting the visual continuity of the park towards the east.

The design of the two towers is less complex than it seems. They consist of two vertical volumes, rising straight on all sides. They are slimmer when viewed from and towards the sea, because the narrower facades face in these directions, and they look broader, therefore, when viewed perpendicular to the coast.

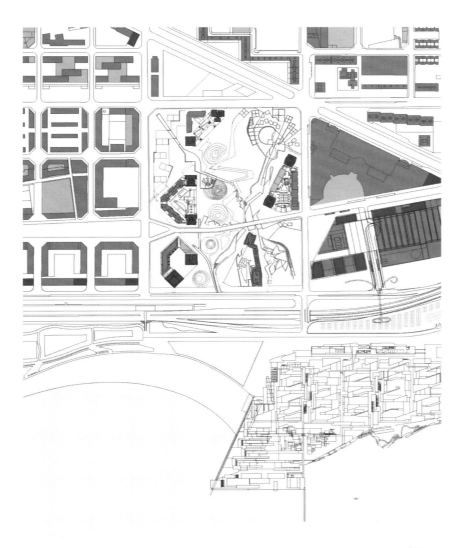

The interior volume is organized around a solid central core, which contains the vertical communications system, and a peripheral, broken line of concrete screens that define the enclosed area. The variations, floor by floor, in the depths of the terraces that surround the closed space, shape the external appearance by creating strategic setbacks, sloping surfaces, overhangs and sharp edges, not strictly vertical.

The exterior facade is formed by continuous horizontal bands of screen-printed plate glass. This protects the terraces from the wind and excessive exposure to sunlight, and filters and clarifies the perception of the interior space, also finished with a covering of tinted glass.

Depending on the incident light and the viewer's position, the color of the various planes of the facade varies during the day from transparent to opaque, offering varying images and shades of color, brightness and transparency.

The building is intended to be understood as a single volume, dynamic and complex, folded and stretched over the ground, and rising sinuously into the sky.

Especially when observed from close-up, when the surfaces are more apparent, and in the views of the narrower facades of the towers from the beach and from Avinguda Diagonal, the building reveals its expressive qualities and establishes an interplay with the organic language of the park that provides formal mutual support.

site plan 总平面图

Illa del Mar 大厦是巴塞罗那城市建设的一部分。该大厦是个远离城嚣的独立式建筑，坐落在 Diagonal Mar 城市公园的临海一侧，最近的地标就是公园本身，街道界定了公园的边界和其余散布在公园周围的建筑。

Diagonal Mar 是一个与众不同的项目，似乎远离城市网格。开放空间与公园面向大海是项目设计需考虑的基本要素。

最初，根据该地区的城市规划，Illa del Mar 被设计为三座高耸的大厦。最显眼的建筑会多或少地占用公园边缘面向大海的中心地块。

巴塞罗那市政厅接受了我们的建议，将三塔变为双塔：一个是较低的塔楼，坐落在 Avinguda Diagonal 的垂直线上，与现存建筑要协调一致；另一个是较高的塔楼，坐落在海滨，朝向公园的尽头，享有从公园到海边的开阔视野。双塔连成一线，较低的塔楼不会影响公园向东的视觉连续性，使整个规划整体感较强。

双塔的设计并不复杂。两座塔楼都是笔直高耸的体量，从海边看，它们都较为纤细，因为较窄的立面面向大海，从垂直于海岸的方向看，它们又很开阔。

内部空间围绕一个中央核心展开布局，其中有垂直交流系统，外围的折线形混凝土屏围合出一个封闭区域。围绕封闭空间的露台深度逐层改变，通过策略性的后缩、倾斜的表面、凸出而尖锐的边缘形成了建筑的外观。

外立面由连续水平的丝网印刷玻璃板构成。这可以保护露台免受风吹日晒，使室内空间的视野更清晰；其顶部覆盖着有色玻璃。

根据入射光线和参观者的位置，立面不同颜色的面板一天之中在透明和不透明之间不断变化，形成不同的形象、阴影、亮度和透明度。

建筑试图展现一种动态而复杂的独立体量，在地面上曲折伸展，直耸云霄。

对建筑进行特写时，其立面形象更加突出。从海滩和 Avinguda Diagonal 观察大厦狭窄的立面时，建筑形象更加传神，并与公园形成有机的互动。

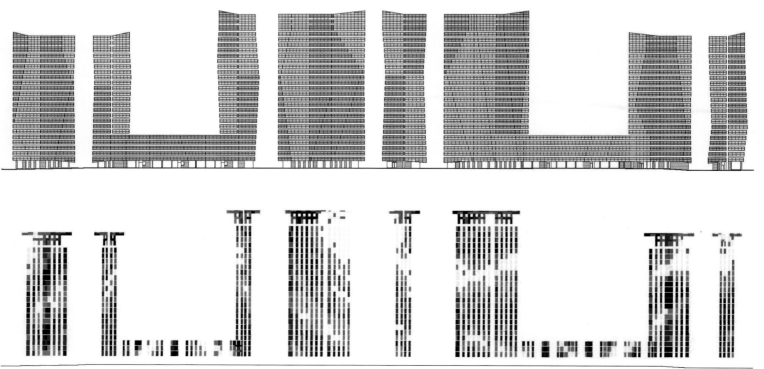

Elevation 立面图

Plan 平面图

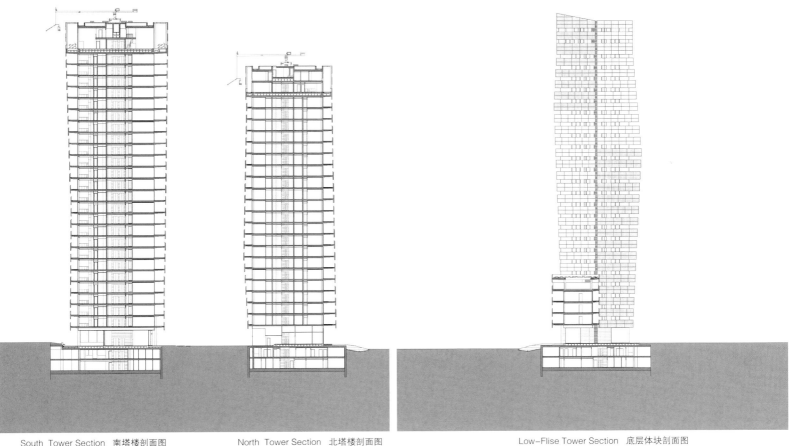

South Tower Section　南塔楼剖面图　　　　　North Tower Section　北塔楼剖面图　　　　　Low-Flise Tower Section　底层体块剖面图

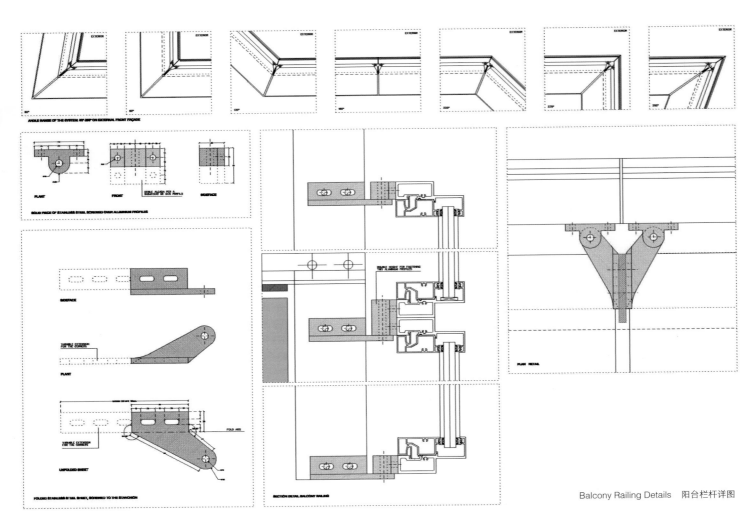

Balcony Railing Details   阳台栏杆详图

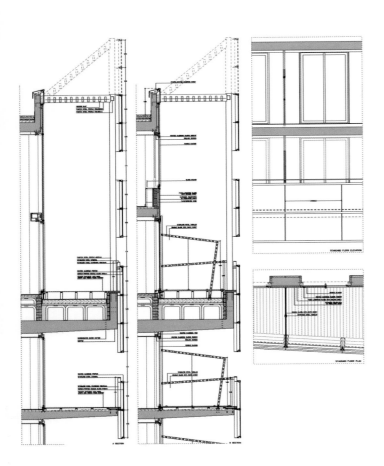

Facade Section Details   立面剖面详图

办公建筑　OFFICE BUILDING

GRANITE TOWER
花岗岩大厦

Architects: Christian de Portzamparc
Client: GALYBET (Sub-company of the group Société Générale)
Location: La Défense, Paris, France
Area: 69,000 m²

设计机构：Christian de Portzamparc
客户：GALYBET（兴业银行集团的子公司）
项目地址：法国巴黎拉德芳斯
面积：69 000 平方米

This tower, which is the new building of the whole headquarters of the French bank Société Générale, is designed as a major feature at the western extremity of Paris, La Défense District.

Tight into a narrow, triangular site, the tower forms a tall and slender triangular prism like a prow marking the start of the Valmy sector. Linked to the existing cylindrical towers of the Société Générale designed in the 1990s by Michel Andrault and Pierre Parat, the triangular tower conveys a dual image: that of an independent building on the Paris, La Défense skyline and that of a building linked to the two existing towers which unifies the bank's block. A dihedral structure forms the front of the complex while prism-shaped flanks along the back preserve the light and views of the existing towers. These forms provide office space of optimum quality that brings together functionality, flexibility and comfort.

The inclined summit of the tower accentuates its stab and its presence in this prime location in the landscape of Paris - La Défense and Nanterre.

The backfaces oblique is a frequent theme in the towers of Christian de Portzamparc, such as Lille Tower, LVMH Tower, Zeil project in Frankfurt.

这座大厦是法国兴业银行的新总部，是巴黎西部拉德芳斯区的特征建筑。

受限于一个狭窄的三角形地块，大厦采用了高而纤细的三棱柱形式，像瓦尔米区的船首标记。建筑与由 Michel Andrault 和 Pierre Parat 于 20 世纪 90 年代设计的圆柱形的兴业银行原办公楼相接，三棱柱的大厦表现出双重形象：巴黎拉德芳斯区的一座独立建筑；与原有办公楼相连将银行区统一成一个整体。建筑正面采用双平面结构，侧面采用棱柱结构，不会影响原有办公楼的光线和视野。这种形式优化了办公空间且集功能性、灵活性和舒适性于一体。

大厦倾斜的顶层强调了其在拉德芳斯和楠泰尔景观中的重要位置。

背面倾斜是 Christian de Portzamparc 所设计建筑的常见特征，如 Lille 大厦、LVMH 大厦和法兰克福市的 Zeil 项目等。

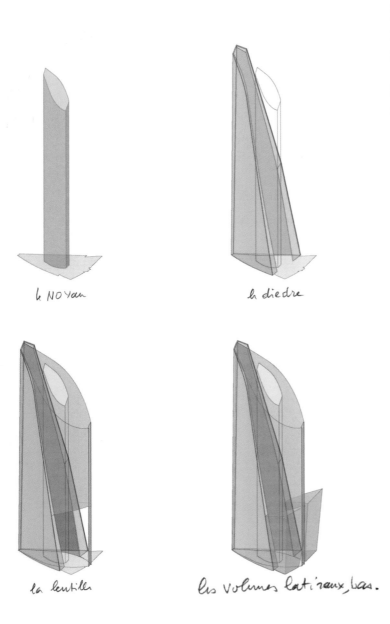

le NOyau

le diedre

la lentille

les volumes latiraux, bas.

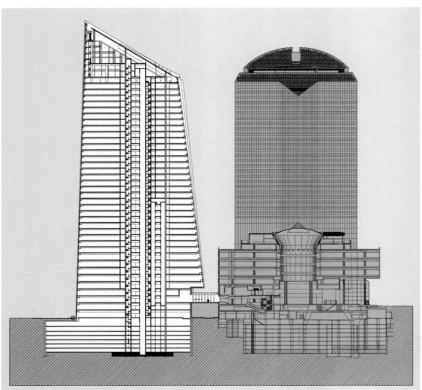

Elevation 立面图

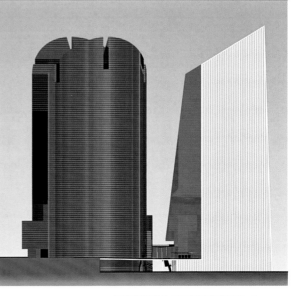

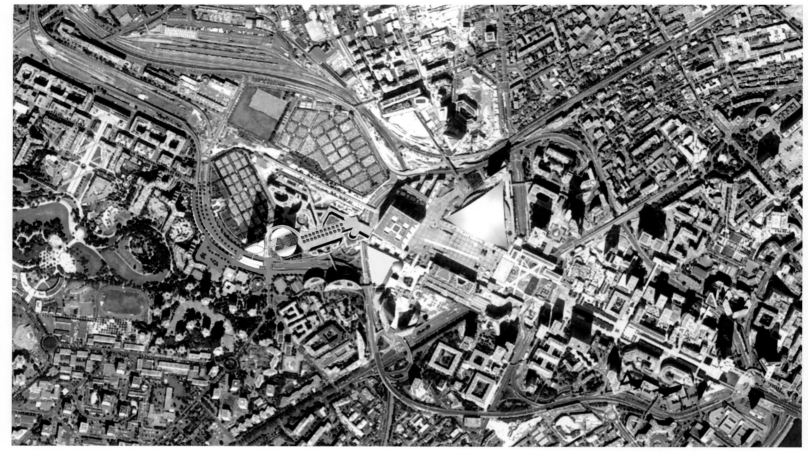

Aerial View 鸟瞰图

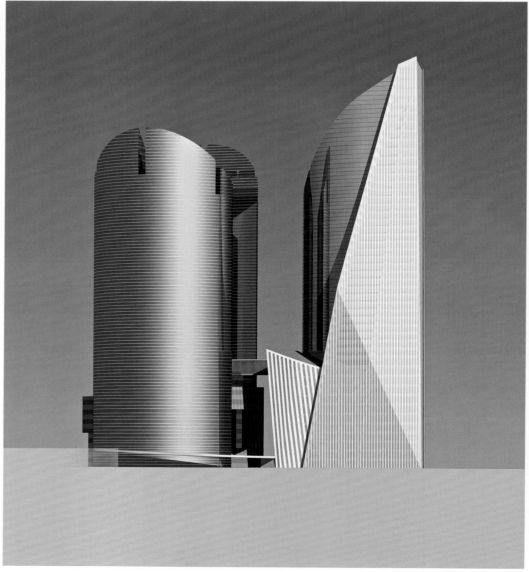

Modelling 模型图

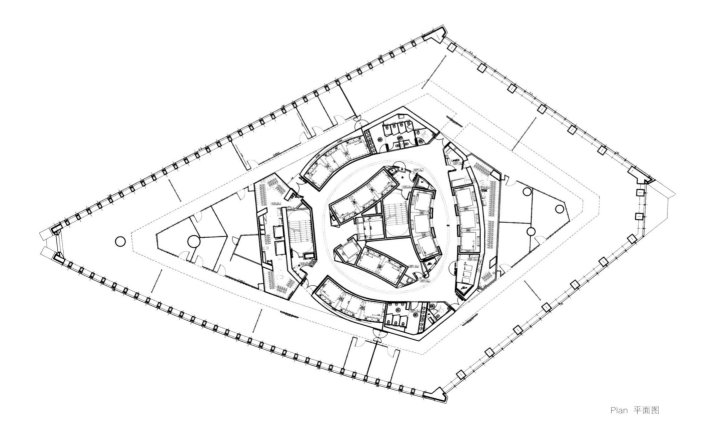

Plan 平面图

NEW HEADQUARTERS BUILDING OF THE SHENZHEN STOCK EXCHANGE
深圳证券交易所新总部大楼

Architects: OMA
Client: Shenzhen Stock Exchange
Partners in Charge: Rem Koolhaas, David Gianotten
Partner: Ellen van Loon
Assistant Architect: Michael Kokora
Location: Shenzhen, China
Site Area: 39,000 m²
Photographer: Philippe Ruault (OMA)

设计机构：OMA
客户：深圳证券交易所
主管合伙人：Rem Koolhaas, David Gianotten
合伙人：Ellen van Loon
助理建筑师：Michael Kokora
项目地址：中国深圳
占地面积：39 000 平方米
摄影：Philippe Ruault（OMA）

For millennia, the solid building stands on a solid base; it is an image that has survived modernity. Typically, the base anchors a structure and connects it emphatically to the ground. The essence of the stock market is speculation: it is based on capital, not gravity. In the case of Shenzhen's almost virtual stock market, the role of symbolism exceeds that of the program — it is a building that has to represent the stock market, more than physically accommodate it. It is not a trading arena with offices, but an office with virtual organs that suggests and illustrates the process of the market.

All of these factors suggest an architectural invention: our project is a building with a floating base. As if it is lifted by the same speculative euphoria that drives the market, the former base has crept up the tower to become a raised platform. Lifting the base in the air vastly increases its exposure; in its elevated position, it can "broadcast" the activities of the stock market to the entire city. The space liberated on the ground can be used as a covered urban plaza, large enough to accommodate public events.

The Shenzhen Stock Exchange(SZSE) — which will rise to 246m — is planned as a financial centre with civic meaning, located in a new public square at the meeting point of the north-south axis between Mount Lianhua and Binhe Boulevard, and the east-west axis of Shennan Road, Shenzhen's main artery.

The raised base of the SZSE is a three-storey cantilevering platform floating 36m above the ground, with a floor area of 48,000 m² and an accessible roof garden. The platform and the lower tower contain the dedicated stock exchange functions, including an international conference center, exhibition spaces, listing hall, and market watching department. The tower is flanked by two atria — a void connecting the ground directly with the trading floor. Staff enter to the west atrium, the public to the east atrium. Surrounding the west atrium is a 20,000 m² base of commercial facilities. SZSE executive offices are located just above the raised podium leaving the uppermost floors free for rental offices and a VIP club.

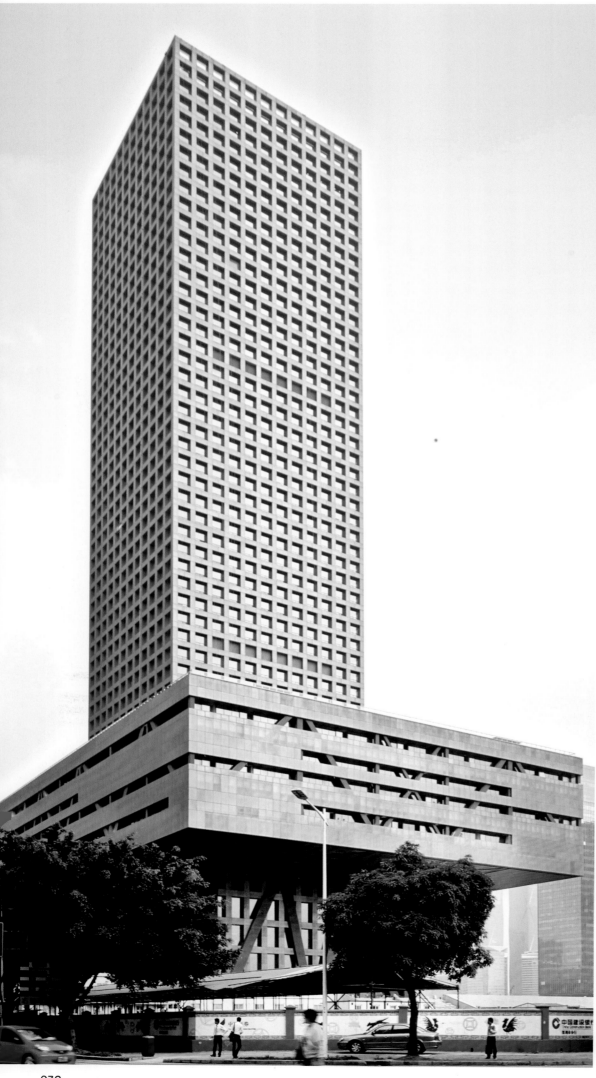

The generic rectangular form of the tower obediently follows the surrounding homogenous towers, but the SZSE's facade is different. The tower's structure is a robust exoskeletal grid overlaid with a patterned glass skin — the first time this type of glass has been used for an exterior in China. The patterned glass reveals the detail and complexity of construction while creating a mysterious crystalline effect as the tower responds to light: sparkling during bright sunshine, mute on an overcast day, radiant at dusk, glimmering during rain, and glowing at night.

The SZSE building is designed to be one of the first 3-star green rated buildings in China. It utilizes passive shading through recessed openings that form a "deep" facade reducing the amount of solar heat gain entering the building, improving natural day lighting while reducing the energy consumption. Intelligent lighting systems shut down the interior day lighting when spaces are not in use. Rainwater collection systems are used and the landscape design is permeable to collect water locally and reduce run-off.

The 200,000 m² building is scheduled for completion in 2013. The three-storey floating podium was built with 27,000 tons of steel. A single joint in the supporting truss work weighs as much as 172 tons. OMA oversees the ambitious engineering and detailed design by working in an office with the client directly on the construction site — an unusual practice for foreign architects working in China. OMA's team is led by Associate Michael Kokora, with general project management by Partner David Gianotten.

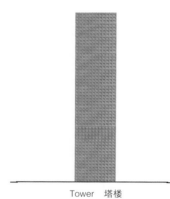

Tower　塔楼

Tower +Base　塔楼＋基座

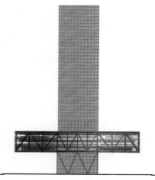

Lifted！　提升！

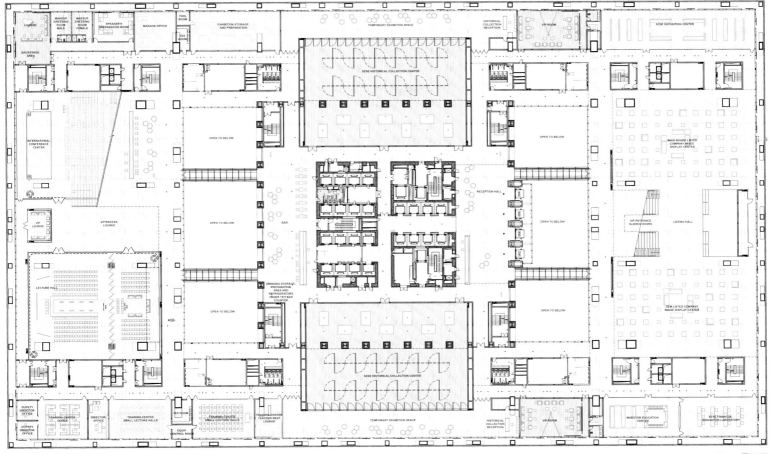

Plan 平面图

几千年来，建筑均矗立在坚实的地基上。这作为一种形象在现代的建筑中得以保存。通常，地基用来固定结构，使它们与大地紧密相连。股市的本质是投机，它是以资本为基础的，而不是重力。对于深圳大多数虚拟股票市场来说，此项目具有象征意义，它代表着股票市场，而不仅仅是股票市场的建筑本身。它不仅是一个可以办公的交易场所，也是一个暗示和描述股市流程的虚拟机构。

所有这些因素表明一种建筑创新：这个项目建在一个悬浮型的基座上，就好像驱使着市场的投机陶醉感骤然上升一样，先前的基座爬上塔楼变成一个升起的平台。将基座升到空中增加了其曝光度，在其升起的位置，可以向整个城市"广播"股票市场的动态。地面被解放出来的空间可以作为一个带顶的城市广场，大到足以举行公共活动。

246 米高的深圳证券交易所被规划为一个有着市民意义的金融中心，坐落在莲花山和滨河大道南北轴线与深圳的主动脉——深南路东西轴线交点处的公共广场上。

深交所三层的悬臂基座悬浮于距地面 36 米处，建筑面积 48 000 平方米，还有一个屋顶花园。平台和下部楼层为证券交易专用场所，包括国际会议中心、展览空间、上市大厅和市场监察部门。大厦两侧是中庭，将首层空间与交易楼层直接相连。工作人员走西侧中庭，公众走东侧中庭。围绕西侧中庭的是一个 20 000 平方米商业空间的底座。深交所行政办公室位于抬升的基座之上，最高层是写字楼出租区和 VIP 俱乐部。

大厦一般采用矩形结构，与周围的建筑类似，但深交所大厦的立面则有所不同。大厦采用覆盖有压花玻璃的外骨架网格结构，这种玻璃结构还是第一次在中国用于外立面。压花玻璃展示了建筑的细部和复杂性，创造出神奇的水晶效果：在明媚的阳光下闪闪发光，在阴天则沉寂着，在黄昏映照着夕阳，在雨天闪烁着点点微光，在夜间则散发出绚丽的灯光。

此建筑旨在成为中国第一座三星级绿色建筑。它采用凹形开口形成被动阴影，形成一个进深立面以减少太阳热量进入建筑，改善自然采光，减少能源消耗。智能照明系统在空间闲置期间自动关闭室内照明。雨水收集系统及可渗透景观设计方案用来局部收集雨水，减少径流。

这个 200 000 平方米的建筑于 2013 年完工。三层悬浮型平台使用了 27 000 吨钢材。支撑桁架的单缝接头重达 172 吨。OMA 与客户在施工现场一起进行了细部设计，这是这个外国建筑公司在中国不同寻常的实践。OMA 团队由 Michael Kokora 领导，总项目管理由 David Gianotten 负责。

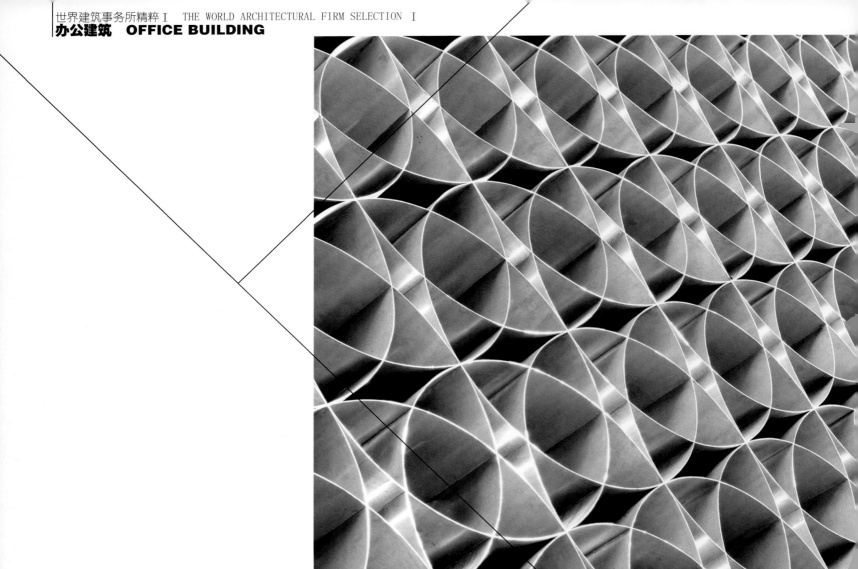

BMCE BANK BRANCHES
BMCE 银行支行

Architects: Foster + Partners
Client: BMCE Bank
Associate Architects: Amine Mekouar, Karim Rouissi (Empreinte d'Architecte)
Landscape Architect: Michel Desvigne
Lighting Consultant: George Sexton Associates
Location: Rabat, Casablanca and Fes, Morocco
Site Area: Rabat Branch – 960m², Casablanca Branch – 500m²,
Fes Branch– 1,280m²
Photographer: Nigel Young

设计机构：Foster + Partners
客户：BMCE 银行
合作机构：Amine Mekouar, Karim Rouissi
(Empreinte d'Architecte)
景观设计：Michel Desvigne
灯光顾问：George Sexton Associates
项目地址：摩洛哥拉巴特、卡萨布兰卡和
非斯
占地面积：拉巴特支行 960 平方米，
卡萨布兰卡支行 500 平方米，非
斯支行 1 280 平方米
摄影：Nigel Young

Banque Marocaine du Commerce Exterieur (BMCE) is one of Morocco's leading banks. Prompted by a desire to transform the experience of high-street banking for its customers, the bank commissioned a series of flagship branches. Their design follows a modular, thematic approach, with variations in scale and landscaping in response to the different locations. While the branches in Casablanca and Rabat reflect their compact sites in the financial and civic centres on Morocco's coast, the Fez branch has subtle details that express the city's artisan heritage.

Each of the banks is entered through a colonnade and topped by a dome. The soffit of the dome is rendered in tadelakt, a local plaster technique, and the exterior is clad in zellige, a traditional ceramic tile. Structurally, the buildings comprise a reinforced concrete frame, with bays repeated on a modular grid. The bays are enclosed by glazed panels with deep screens to provide shade and security. The screens are made from a low-iron stainless steel, which is designed to remain cool to the touch, and follow a geometric design based on Islamic patterns. The combination of screens and stone pillars gives the facades the appearance of solidity, in keeping with the often decoratively carved walls prevalent in Morocco's refined Arabic style. While the building envelope relates to the regional vernacular, the interior is contemporary. The dome form sweeps down into the banking hall to create a sculptural curved bench — a distinctive feature of each branch — which varies in width according to the building's size.

A simple functional arrangement divides the rectangular footprint equally between the banking hall and support areas. The modular grid places landscaped pools, with planting in recessed stone rills, around the exterior of the building. The branches are designed to be energy efficient, combining modern technology with centuries-old methods of passive climate control. One example is the use of "earth-tube" cooling, in which fresh air is drawn into a pipe that encircles the structure below ground and is chilled naturally before being released into the building's ventilation system.

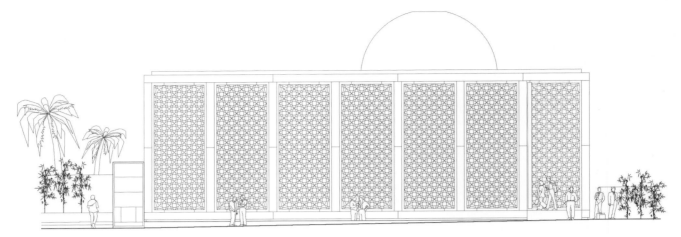

Southeast Elevation　东南立面图

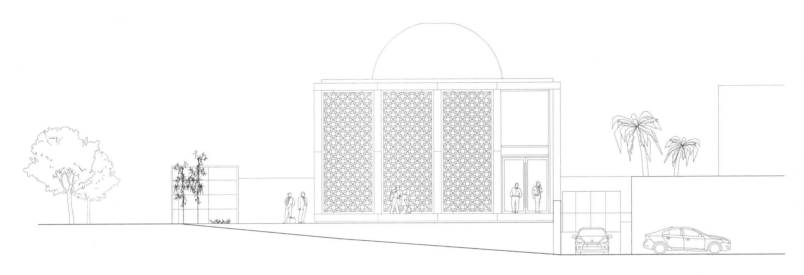

Northeast Elevation　东北立面图

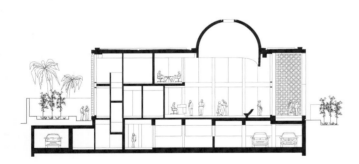

Long Section　纵向剖面图

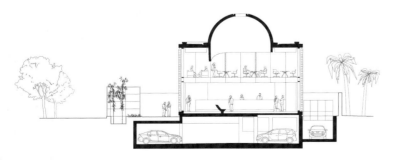

Short Section　横向剖面图

　　BMCE 银行是摩洛哥的主要银行之一。为了改变客户在商业街银行的业务体验，BMCE 新建了一系列的旗舰支行。这些支行的设计遵循模块化、主题化的方法，在不同地点具有不同的规模和景观。卡萨布兰卡和拉巴特的支行反映了它们在摩洛哥海滨金融和市民中心的紧凑之感，非斯支行的微妙细节则体现了城市的文化遗产。

　　各支行的入口都有一个柱廊，顶部是穹顶。穹顶的拱腹处以 tadelakt（当地的一种石膏材料）打底，其外部用 zellige（一种传统瓷砖）覆盖。在结构上，这些建筑为钢筋混凝土框架，在单元网格上有多处飘窗。飘窗由玻璃板和深屏封闭，具有遮阳和保护功能。屏风由低铁不锈钢制成，摸起来手感凉爽，其上的几何图案富有伊斯兰风情。屏风和石柱相结合，使外立面展现出坚实的外观，与在摩洛哥盛行的具有阿拉伯风情的雕刻装饰墙相协调。建筑外壳与当地方言相联系，内部则采用当代的建筑风格。穹顶向下延伸到银行大厅，创造了一个雕塑般的曲线形长凳，这是每个支行的特色，其根据建筑的尺度具有不同的宽度。

　　简单的功能布局将银行大厅和辅助区之间的矩形区域等分。模块化的网格设有景观池，建筑外围有小溪环绕。这些支行的设计均采用了节能技术，同时将现代技术与被动适应气候的古老方法相结合，如采用地下管冷却，新鲜空气通过地下的管道引入室内，在进入室内通风系统之前已被自然冷却。

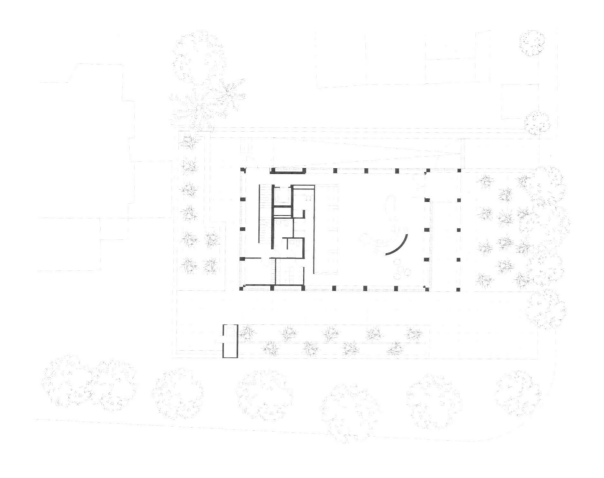

Ground Floor Plan 底层平面图

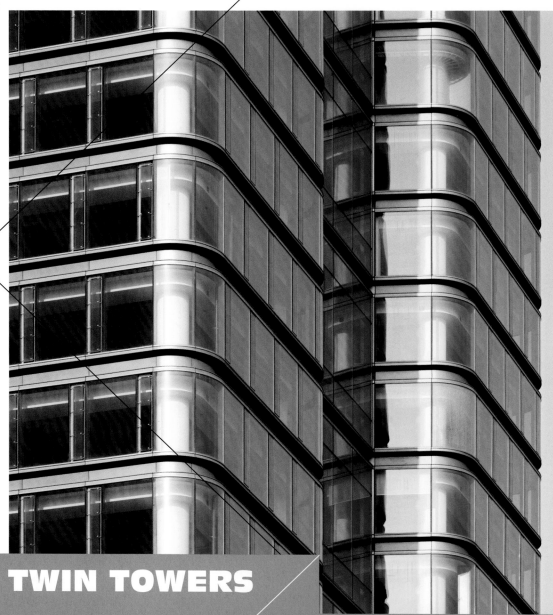

HUALIAN TWIN TOWERS
华联双子塔

Architects: GMP(von Gerkan, Marg and Partners)Architects
Client: Hangzhou Union Developing Group of China
Design Partner: Meinhard von Gerkan, Nikolaus Goetze
Lighting Design: Schlotfeldt Licht, Berlin
Chinese Partner : Zhejiang Chengjian Construction Group Co., Ltd.
Location: Hangzhou, China
Building Area: 95,400 m²
Photographer: Hans Georg Esch

设计机构：GMP 建筑师事务所
客户：中国杭州国联发展集团
合作设计：Meinhard von Gerkan, Nikolaus Goetze
灯光设计：Schlotfeldt Licht, Berlin
中方机构：浙江城建建设集团有限公司
项目地址：中国杭州
建筑面积：95 400 平方米
摄影：Hans Georg Esch

The buildings is in an unusual situation in a new part of Hangzhou directly on the banks of the broad River Qiantang. The twin towers have square ground plans, and are joined by a three-story base. The towers are skewed relative to each other, to allow the optimum views from all offices and reduce visual intrusion. Two-storey winter gardens are integrated into the office floors on two opposite sides of the building as thermal buffers and to reduce energy emissions. Large supply and waste air apertures provide ventilation and prevent the offices overheating in summer. The double facades of the towers consist of fixed glazing and opening windows, with the floor-to-ceiling-height, unitized fixed glazing being made of neutral-color, high-efficiency insulating glass. The opening windows on the inside provide natural ventilation. Even in rain or storms, the office premises can be ventilated by arranging the windows, which considerably improves convenience.

Project Manager: Volkmar Sievers
Project Team: Wiebke Meyenburg, Diana Spanier, Ulrich Rösler, Simone Nentwig, Alexandra Kühne, Barbara Henke

项目经理：Volkmar Sievers
项目团队：Wiebke Meyenburg, Diana Spanier, Ulrich Rösler, Simone Nentwig, Alexandra Kühne, Barbara Henke

　　该项目位于杭州钱塘江畔一个显著位置。双塔占据一个方形地块，由一个三层的基座相连。塔身彼此相对倾斜，使所有办公室都可拥有最佳视野并减少视觉干扰。在建筑的两端，两层的冬季花园融入办公楼层，可起到热缓冲器的作用并减少能源损失。大量的进气孔和排气孔加强了通风，防止夏季温度过高。建筑的双层立面上镶嵌有固定式玻璃开放落地窗，落地窗采用中性色高效真空玻璃。开放式落地窗保证了自然通风，即使在狂风暴雨的天气，这种窗户也能方便地提供良好的通风。

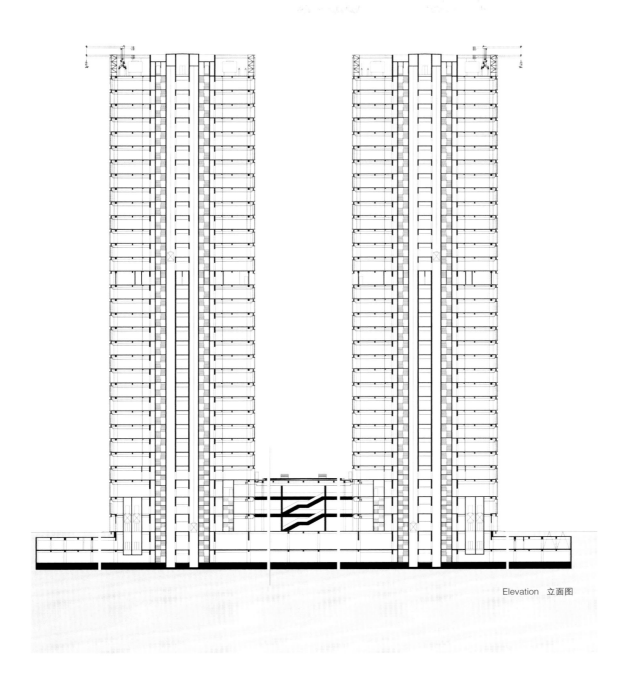

Elevation 立面图

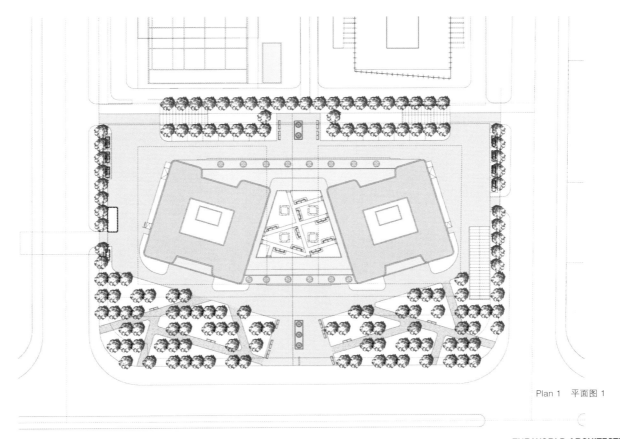

Plan 1 平面图 1

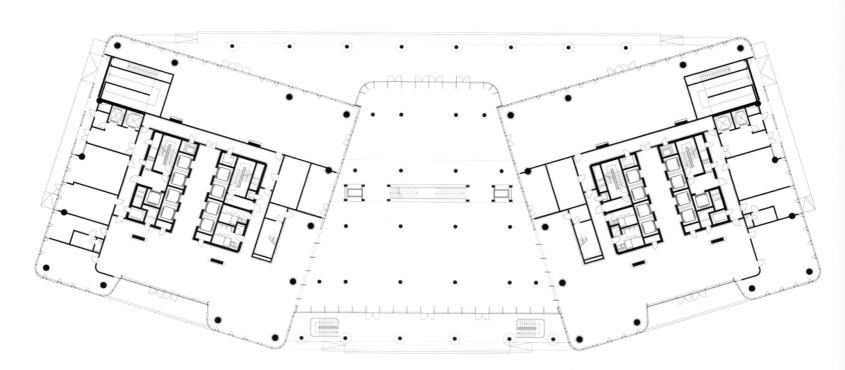

Plan 2 平面图 2

THE BUILDING LA LIBERTÉ
LA LIBERTÉ 大厦

Architects: Dominique Perrault Architecture, Paris, France
Client: Christelijke Woningstichting Patrimonium,The Netherlands
Location: Groningen, The Netherlands
Site Area: 10,230 m²
Building Area: 23,400 m²
Tower A: 80m high / 22 floors / 13,000 m²
Tower B: 40m high / 10 floors / 5,800 m²
Photographer: Prima Focus / Mark Sekuur / DPA / Adagp

设计机构：法国巴黎多米尼克·佩罗建筑师事务所
客户：荷兰 Christelijke Woningstichting Patrimonium
项目地址：荷兰格罗宁根
占地面积：10 230 平方米
建筑面积：23 400 平方米
A 座：80 米高 /22 层 /
13 000 平方米
B 座：40 米高 /10 层 /
5 800 平方米
摄影：Prima Focus / Mark
Sekuur / DPA / Adagp

Associate Architects: Oving Architekten, Groningen
Engineering Architects: Dijkhuis
Groningen(structures),
Zonderman B.V. (electrical engineering),
Feenestra (heating and ventilation systems),
Gevekebouw B.V. (general contractor)

合作机构：Oving Architekten, Groningen
工程机构：Dijkhuis Groningen（结构工程），
Zonderman B.V.（电气工程），
Feenestra（暖通工程），
Gevekebouw B.V.（总承包方）

"La Liberté", social housing and office building, has been realized within a constant collaboration among the client, the architects and the City of Groningen.

The project is a strong symbol of the urban renewal and the social development of the city. New signal in the landscape, "La Liberté" was an immediate success: all the apartments have been already rented!

Groningen, a young city, numbered among the 10 important cities of the Netherlands, has been developing since 2004 a housing campaign called "The Intense City". By avoiding the suburbanization, protecting the rural areas and guaranteeing quality of urban spaces, the municipality wishes to keep the city compact and alive.

Although strengthened by this new urban vision for Groningen, the city has launched afterwards the Ring Zuid Groningen initiative, to develop more particularly the neighbourhoods lining the main highway of the country, Weg der Verenidge Naties, located in the South of Groningen.

This new strategy, encouraged by the Ministry of Transports, the province and the city of Groningen, is a major issue not in the urban and social renewal of the city, but also in the environmental area.

First of all, this highway, acting as a real urban border, will be gradually trenched and "divided up" in order to improve and intensify the urban circulations and links. Around the highway, the city has recommended high rise architectures in order to create and preserve green spaces at the bottom of the constructions. The preservation of those green spaces and their connections will allow an ecologic lane, opened to everyone and guaranteeing a new bio diversity.

In parallel, the city has started to work with different property developers for the realisation of office and/or social housing buildings and to appoint international architects. Several projects are under construction and in particular some offices by UN Studio and some housing by Mecanoo.

So, in 2007, the Christelijke Woningstichting Patrimonium Company has ordered Dominique Perrault to the construction of two mix-used buildings, rental social housing and offices, in the south-west of the city.

Located at the outskirts of Groningen, the parcel is surrounded by typical post war social housing blocks. Dominique Perrault will take advantage of the urban context to create the volumes of "La Liberté". As the request was to build high up, the architect leans on this horizontality to create a crescendo of volumes.

The project comprises the construction of two square plan buildings: one tower of approximately 80 meters (Tower A), and another of 40 meters high (Tower B).

The buildings are both made up of a platform, entirely in glass, independent, with the same height (R+2) and accommodating the offices. As they are not taller than the nearby blocks, the platforms respect and extend their horizontality.

Then, two blocks, with different heights, seem to be floating above the platforms and accommodate the housings. Here the architect plays with the volumes: actually the tower A is made up of two volumes of housing with equivalent proportions, and slightly shifted. It seems that the architect has piled up different volumes, one on the others: one volume of offices and one volume of housing for the tower B, one volume of offices and two shifted volumes of housing for the tower A.

The housing blocks are hanging above the offices thanks to a 5 meters high terrace. In this "in-between", only a core, sheltering the common spaces, brings an easy transition between the private spaces and the offices.

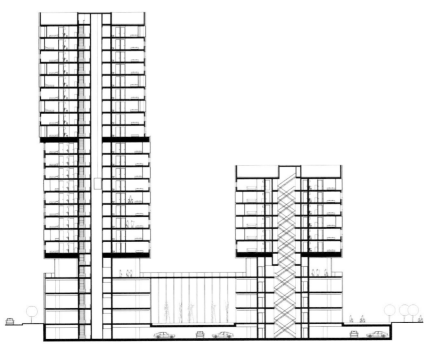

Seltion　剖面图

At last, a footbridge, located at the same level and opened to the users, links up the both towers.

Through the treatment of the facades, Dominique Perrault creates a real dialogue between the both buildings and between the project and its urban environment. Whatever the viewpoint, the facades of the three housing blocks never offer the same treatment of colours — black, grey and white. These shades of colour strengthen the stack impression and energize the city skyline.

Moreover some polished steel panels are placed perpendicularly to the facades, which punctuate the facades and multiply the perceptions of the building.

Dominique Perrault delivers the social housings of quality, offering substantial dimensions and spaces. Each level has 5 apartments; all the apartments have 3 rooms and a wide average surface area.

The Dutch law enforces a totally deaf facade for the apartments facing the highway. This housing can't have any opening window and there is no ventilation. Dominique Perrault plays with this imposition by articulating these apartments around glass inner loggias. These loggias, totally mobile and equipped with sophisticated filters, work as a real protection against noise pollution.

Finally the polished steel panels, set up at the loggias level, capture natural light and reflect it towards the apartments. These mirrors also offer to the residents an alternative outdoor vision.

"La Liberté" counts 120 apartments, split up in 15 different types with about 40 variants.

So the high rise construction allows free spaces at the bottom of the

Plan du rez-de-chaussée

0 5 10 20   50

buildings. At the North of the building, and to answer to the order, Dominique Perrault lays out the ecologic garden designed to develop the local flora and fauna. Opened to everyone, it will be integrated into the city ecologic way. A "meditative" garden will be created at the South of the buildings. It will be a space offered and opened to everyone.

Dominique Perrault introduces here many levels of dialogue: between the project and its urban environment; between the both towers through their architectural treatment; between the public and the private spaces through easy transitions allowing the exchange between the residents and the people working in the offices.

Dominique Perrault realizes a building opened to the city: through this artful manipulation of differences in scale these buildings have been designed to become a new landmark signal, creating a new urban landscape.

"自由居"——"La Liberté"（社会住房兼办公楼）在甲方、建筑师及格罗宁根市政府始终如一的通力协作下竣工了。

该项目是城市改造和城市社会发展强有力的象征。"自由居"作为当地景观的新标志，一经建成就大受欢迎：目前所有的公寓都已租出！

格罗宁根是一座年轻的城市，为荷兰十大主要城市之一，自 2004 年以来就一直在发展一项名为"密集城市"的住宅运动。市政府希望通过避免郊区化，保护乡村地区不被侵吞，确保城市空间的质量，使城市保持紧凑的状态与旺盛的活力。

尽管这个新的城市项目提升了格罗宁根的整体环境，但是市政府仍然相继开发了几个重点项目：格罗宁根 Ring Zuid 项目，连接国家主要高速公路，开发更多更特别的居民区；Weg der Verenidge Naties 项目，位于格罗宁根的南部。

在荷兰交通部、格罗宁根省和格罗宁根市的鼓励下，这个新设计方案不仅成为城市改造与社会改革的主要议题，在环境领域亦是如此。

首先，为了提升城市交通能力，增加交通路线的数量，这条高速公路作为真正的城市边界将逐渐被"瓜分"。

在高速公路的周围都是城市高层建筑,以此来创造和保护建筑底部的绿色空间。保留这些绿色空间及其之间的通道,将形成对所有人开放的生态路径,确保了新的生物多样性。

与此同时,市政府开始与多家房地产开发商合作开发办公楼和/或社会住房,并将设计任务委托给一些国际建筑师。许多项目正在建造之中,例如由 UN 工作室设计的办公楼以及由 Mecanoo 建筑事务所设计的住宅楼。

因此,2007 年,Christelijke Woningstichting Patrimonium 公司邀请多米尼克·佩罗在城市的西南区域设计建造两座混合用途建筑,即租赁型社会住房与办公楼。

建筑场地位于格罗宁根郊区,周围都是典型的战后建造的社会住宅区。多米尼克·佩罗将利用城市文脉创造"自由居"的建筑体量。设计要求是建造参天大楼,所以建筑师在地面上设计了一种递增型的体量。

该项目由两座平面布局为正方形的建筑构成:一座大楼约 80 米高(A 座),另一座 40 米高(B 座)。

两个建筑底座均由独立的玻璃结构平台构成,高度相同,内部为办公场所。考虑到与周围环境的关系,它们并未比附近的建筑高,并且扩展和延伸了原有建筑的水平状态。

接下来是两个高度不同的建筑体块,它们似乎漂浮在平台上方,其内部为住宅。在这里,建筑师用这些建筑体量做起了堆积木的游戏:实际上 A 座的两个住宅体量大小一致,彼此略微错开位置,看上去就好像建筑师将各种不同体量的建筑堆积起来。B 座由一个办公楼和一个住宅楼堆成,而 A 座则由一个办公楼和两个错位的住宅楼堆成。

"悬浮"于办公楼上方的住宅体块是通过 5 米高的露台相连的。在连接处仅有一个中心区域作为公共空间,为办公空间与私人空间之间提供了简洁的过渡区域。

最后有一座人行天桥位于两座大楼的相同高度处,起到了连接的作用,面向使用者开放。

多米尼克·佩罗通过对建筑立面的处理,在两座大楼之间以及项目与周围城市环境之间建立了真正的联系。无论从哪个角度看去,三个住宅体块的外观都呈现出黑、白、灰三种不同的颜色。这些颜色增强了堆叠的印象,进一步为城市的天际线增添了活力。

此外,一些抛光钢面板被垂直放置在立面上,不但突出了立面的特色,还丰富了建筑的外观形象。

多米尼克·佩罗设计的这个社会住宅项目品质出众,空间宽阔且实用。每一层都有 5 间公寓,所有的公寓都有 3 个房间,平均使用面积很大。

荷兰法律规定,面对高速公路的公寓立面必须完全采取封闭式设计。这个住宅项目没有一扇能敞开的窗户,因而无法通风。于是多米尼克·佩罗将这些公寓围绕着室内玻璃凉廊建造。这些凉廊可以整体移动,并且配备了精密的过滤器,可真正地保护居民免受噪声干扰。

最后,安装在凉廊上的抛光钢面板可吸收自然光线,将之反射到公寓内部。这些镜子令居民有一种身处户外之感。

"自由居"共有 120 套公寓,分为 15 种户型,细分又有大约 40 种类型。

这个高层建筑在建筑底部保留了自由空间。在建筑北面,为了与整体环境相符,多米尼克·佩罗设计了生态园林,开发当地的动植物群。园林向所有人开放,与城市的生态环境相融合。建筑师将会在建筑的南部建造一座"冥想"花园。这个花园也将会对所有人开放。

多米尼克·佩罗在这个项目中引入了许多层面的建筑对话:项目与城市环境之间的对话;通过对建筑的处理形成的两座大楼之间的对话;公共与私人空间之间的对话。这也让居民与办公室工作的人们之间的交流更加顺畅。

多米尼克·佩罗实现了一个面向城市开放的建筑:通过对建筑规模的巧妙处理,这两座大楼成为一个崭新的城市地标,创造出全新的城市景观。

SPIEGEL HQ
德国明镜周刊办公总部

Architects: Henning Larsen Architects A/S
Designer: Martha Lewis
Project Team: Charlotte Bigom, Daniel Illum-Davis, Esben Wong, Filip
Francati, Henrik Vuust, Katja Brandt, Martha Lewis, Maximilian Müller,
Merete Alder Juul, Mikkel Thorvald Madsen, Peter Gamborg, Peter
Koch, Sarah Kübler, Silke Jögenshaus ,Sonila Thaka
Location: Hamburg, Germany
Area: 50,000 m²

设计机构：Henning Larsen Architects A/S
设计师：Martha Lewis
项目团队：Charlotte Bigom, Daniel Illum–Davis,
Esben Wong, Filip Francati, Henrik Vuust,
Katja Brandt, Martha Lewis, Maximilian Müller,
Merete Alder Juul, Mikkel Thorvald Madsen,
Peter Gamborg, Peter Koch, Sarah Kübler,
Silke Jögenshaus, Sonila Thaka
项目地址：德国汉堡
面积：50 000 平方米

Spiegel is one of the largest media houses in Europe and with its new headquarters
in Hamburg, the Spiegel Group has got one of the most prominent locations in the
new urban quarter of HafenCity by the Elbe.
For more than 40 years, the world-known news magazine, Der Spiegel, has been
located in Brandstwiete of Hamburg, in a building very much characterized by
Verner Panton's colourful pop art design. Spiegel Online and Spiegel TV have been
situated at other locations in the city, but now the three divisions are consolidated
in one building—the new Spiegel Headquarters.
The new headquarters has a prominent location in the inner harbour of Hamburg
and offers a distinctive design with its own unique architectural expression.
Together with the office building, Ericus-Contor, which forms part of the complex,
Spiegel Headquarters rises from a red tile base, above which floats a bright,
transparent building volume of glass, steel and concrete.
Spiegel Headquarters is designed by Henning Larsen Architects and is characterized
by its clear focus on contact points, meeting places and communication region.
In combination with its high architectural quality, clear connection to the urban
context and holistic sustainable design, the project is designed to meet the
needs and requirements of a modern media house. This was also the reason
why the project was selected as winner of the international competition with 13
participants in 2007.
The new headquarters covers 30,000 m² and houses 1,100 employees. The
bright, transparent design supports the work processes in the editorial offices,
documentary department and publishing division and the interior layout provides
room for concentration, communication and dis-semination. The individual floors
are connected by stairs and footbridges rising across the central atrium space. The
large central window of the building, "Fenster zur Stadt", creates an active dialogue
between the activities of the media group and city life.
The fifth floor comprises a café very much inspired by the interior design of
Panton in the previous Spiegel Headquarters from 1969. The original red pop
art environment creates an inspiring, evocative gathering point for Spiegel's
employees. From the outside, it stands out as a red orange field in the facade facing
the city. The vivid Panton colours closely connect to Spiegel's identity and have
been re-used several places in the building.

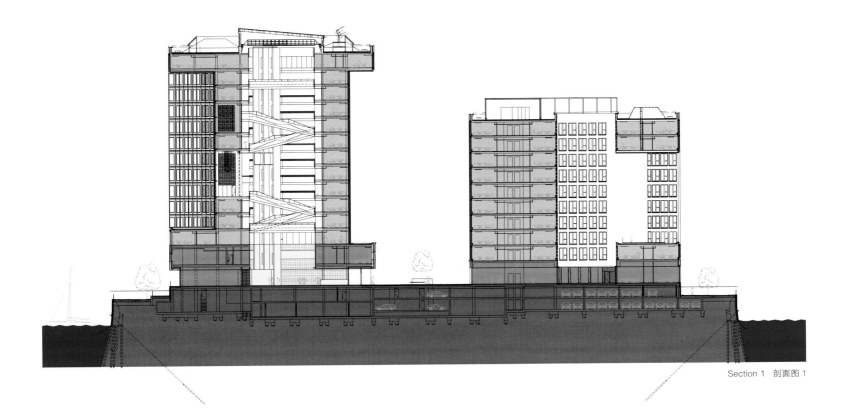

Section 1 剖面图 1

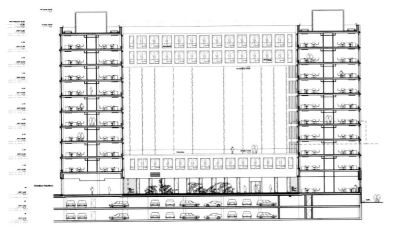

Section 2 剖面图 2

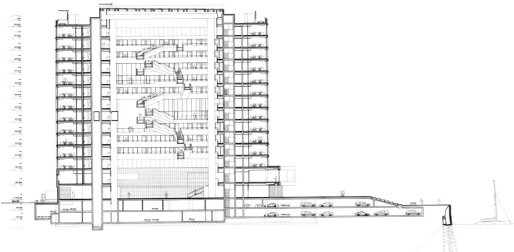

Section 3 剖面图 3

Ericusspitze
Spiegel Section BB, Scale:1:350
Henning Larsen Architects A/S

One of the great qualities of HafenCity is the proximity to the water. It has been an essential focus point in the design process that the new Spiegel Headquarters not only offers the employees an attractive office environment, but also contributes positively to exploit the public urban space. The building connects the old Hamburg and the new urban quarter of HafenCity and creates a vibrant urban space with squares and recreational spaces along the Elbe.

As one of the first and most sustainable buildings in Hamburg, Spiegel Headquarters has been awarded the prestigious award "HafenCity Umweltszeichen Gold"—a forerunner of the German certification standard, DGNB.

　　明镜是欧洲最大的媒体集团之一，其总部在汉堡，明镜集团现已成为汉堡易北河附近港口新城的商业中心区的地标。

　　作为世界闻名的新闻媒体，明镜周刊的办公室40多年来一直位于汉堡Brandstwiete地区一座由Verner Panton设计的充满缤纷色彩的波普艺术大楼里。明镜在线传媒和明镜电视台位于城市的其他地方，但是现今这座新的明镜集团总部大楼可以同时容纳这三个分部。

　　这个新总部所在地在汉堡市内港中具有极大的区域优势，并且其新颖的设计被赋予了独特的建筑语言。作为办公总部综合体的一部分，Ericus-Contor和明镜总部就像是从红色瓷砖地基上面浮起一般，其整体通透，由玻璃、钢筋、混凝土组合而成。

　　明镜周刊总部由Henning Larsen建筑师事务所设计，该建筑的特点体现在联络区、会议空间和交流区域的设计上。结合建筑本身的高质量、与城市环境文脉和整体可持续设计的密切联系，这个项目旨在满足现代媒体集团办公建筑的需求。这也是该项目从2007年的13个参赛项目中脱颖而出获胜的原因。

　　这个办公总部面积为30 000平方米，可以容纳1 100位雇员。明亮透明的办公环境设计有利于编辑部、材料部和发行部工作的开展，室内布局则有利于集中、交流和传播。楼层之间由楼梯和穿过中庭的人行天桥相连。建筑中间的大型窗体被称为"城市之窗"，创造了媒体集团活动和城市生活之间充满活力的对话。

　　建筑第五层是一个咖啡厅，这种布局灵感来自于早前的明镜办公室设计师Panton于1969年的设计。原有的红色波普艺术环境营造出一个充满灵感和凝聚力的场所。从外面看，面向城市立面的橘红色显眼夺目。Panton鲜明的色彩设计成为明镜集团办公室的标志性特征，这种色彩的运用也体现在新的办公总部设计中。

　　港口新城位置的优势之一在于临近水景。这是设计过程中着重考虑的因素之一，明镜周刊新总部不仅为员工提供具有吸引力的办公环境，还有助于开拓城市公共空间。这一办公建筑连接了汉堡旧城和新的港口区域，沿着易北河一带打造出充满活力的城市商业广场和公共娱乐空间。

　　作为汉堡市最环保的建筑之一，明镜周刊总部获得了极负盛名的奖项"HafenCity Umweltszeichen Gold"（德国认证标准DGNB的前身）。

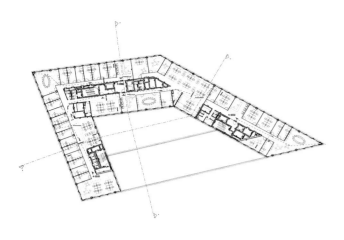

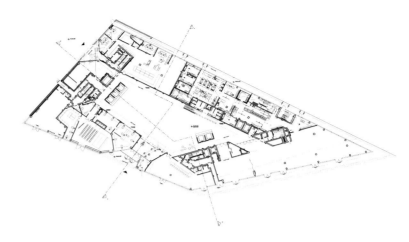

Ericusspitze
Ericus-Contor, 3rd Floor, Scale: 1:350
Henning Larsen Architects A/S

Ericusspitze
Spiegel, Groundfloor, Scale: 1:350
Henning Larsen Architects A/S

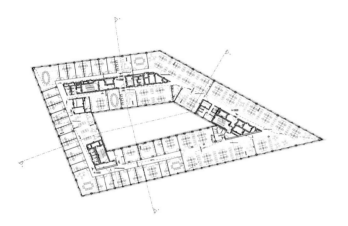

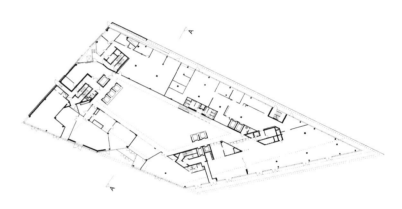

Ericusspitze
Ericus-Contor, 9th Floor, Scale: 1:350
Henning Larsen Architects A/S

Ericusspitze-Spiegel
Level 0
Scale 1:500

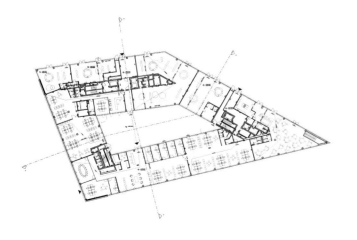

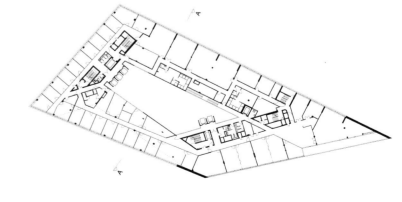

Ericusspitze
Ericus-Contor, Groundfloor, Scale: 1:350
Henning Larsen Architects A/S

Plan 平面图

Ericusspitze-Spiegel
Level 1
M 1:500

555 MISSION STREET
大使街 555 号大厦

Architects: Heller Manus in Collaboration with Kohn Pedersen Fox
Client: Tishman Speyer Properties
Location: San Francisco, USA
Area: 58,062 m²

设计机构：Heller Manus in Collaboration with Kohn Pedersen Fox
客户：Tishman Speyer Properties
项目地址：美国旧金山
面积：58 062 平方米

San Francisco's first LEED Gold Office tower, 555 Mission Street is bounded by Mission Street on the northwest and by Minna Street on the southeast. This 33-storey office building is located to enhance the evolving mid-block open space pattern — where a checkerboard arrangement of buildings has emerged — and will maximize light, air, and public open space opportunities.

The shape of 555 Mission responds to neighbouring structures and the Mission Street corridor by stepping back in the most visible locations, and visually emphasizing verticality and slenderness. This part of the city, adjacent to the Transbay Terminal, is zoned for the maximum height and density allowances in the City's Downtown Plan. 555 Mission will occupy the site adjacent to the existing 101 Second Street adding a major outdoor plaza that forms a connection between the two towers as well as a major mid-block pedestrian link between Market and Mission Streets.

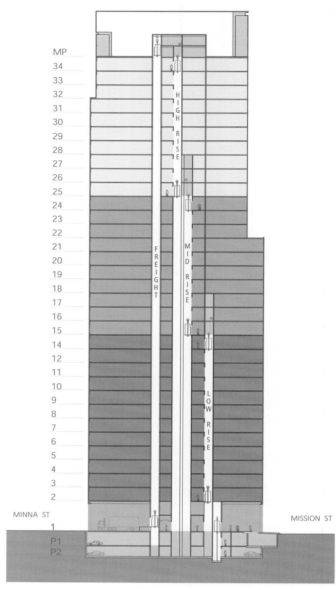

Elevation 立面图

　　大使街 555 号大厦是旧金山第一个获得 LEED 金奖的办公建筑，它西北紧邻大使街，东南邻 Minna 街。这个 33 层的办公楼用于增强道口区间的开放模式，建筑棋盘式的布局可以最大限度地增加光的摄入、空气的流通以及公共空间开放的面积。

　　555 号大厦的形状和周边结构相呼应，其位于大使街最显眼的位置，外形强调垂直性和纤细感。城市的这一部分紧邻 Transbay 枢纽站，在市中心规划中为最高和最密的建筑区域。555 大厦毗邻 101 第二大街，一个大型的户外广场使两座大厦相互联系，同时也在市场和大使街之间创造出一片步行区域。

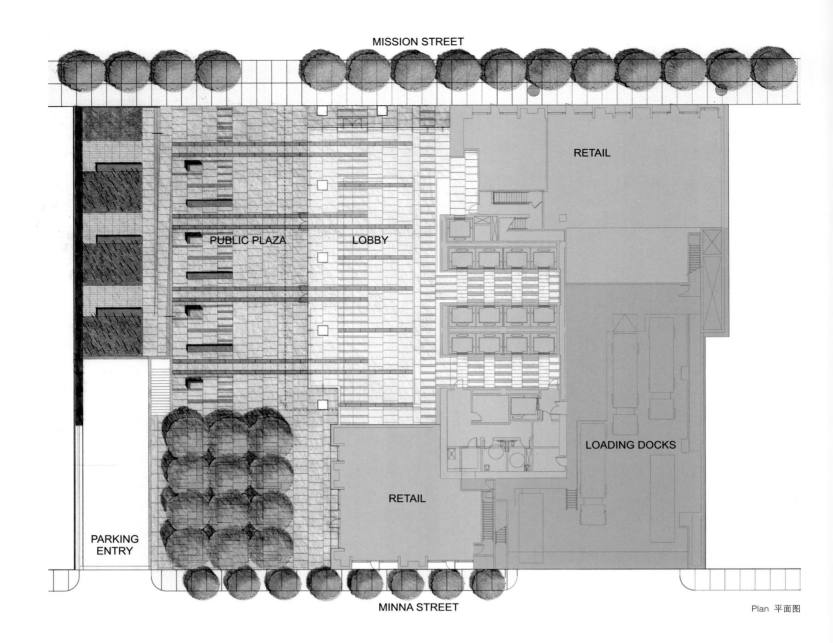

MISSION STREET

RETAIL

PUBLIC PLAZA   LOBBY

LOADING DOCKS

RETAIL

PARKING
ENTRY

MINNA STREET

Plan 平面图

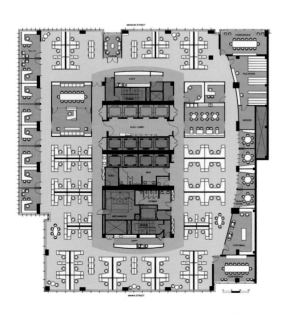 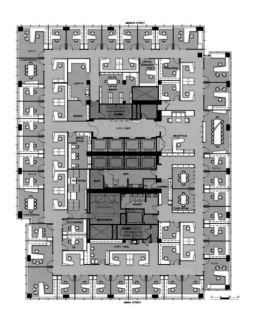 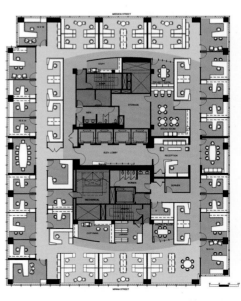

Floor Plan 楼层平面图

KK100
京基 100

Architects: TFP Farrells
Client: Shenzhen King Key Real Estates Development Co., Ltd.
Location: Shenzhen, China
Tower Height: 441.8m
GFA (Tower): 210,000 m^2
GFA (Masterplan): 417,000 m^2

建筑师：TFP Farrells
客户：深圳京基房地产开发有限公司
项目地址：中国深圳
建筑面积（大楼）：210 000 平方米
建筑面积（总体规划）：417 000 平方米
高度：441.8 米

Confirmed by The Council for Tall Buildings and Urban Habitat (CTBUH) as being the tallest building in the world completed in 2011, KK100 is an innovative high density project that takes an entirely new approach to city making. It is situated on the edge of Shenzhen's CBD and sets a new precedent for the successful 21st century transformation of commercial districts into vibrant and enriching environments.

The 100-storey, 441.8-metre tower comprising over 210,000m2 of accommodation is part of the master plan for a 417,000m2 mixed-use development. The development includes five residential buildings and two commercial buildings. The floors of the tower are divided into three major functions. The floors from level 4 to 72 house 173,000m2 of Grade-A office space while the uppermost levels from 75 to 100 are occupied by a 35,000m2 6-star Luxury Hotel complemented with a cathedral-like glazed sky-garden animated by various activities.

Linking all these elements is the podium that is driven by a retail environment that emphasises local identity, excitement, and economic vigour. This will become Shenzhen's most prominent retail address and a destination in its own right. The design provides retail access from every side and there is no dead frontage.

Car parking for the retail mall is located on the third and fourth floors of the podium, meaning that customers can walk straight from their vehicles into the mall on the same level. This greatly increases the convenience of vehicular access compared to traditional underground car parks. Due to this convenience, and the ease of access to the mall for pedestrians on all sides, for occupants of the tower and for residents of the development, the mall will feel like a naturally connected "high street" rather than an enclosed and isolated shopping centre. The mall thus acts as a connector which links all elements of the development and also integrates it well with the neighborhood at street level.

As well as providing social and cultural continuity, KK100 is integrated with the metropolitan transport network, which is crucial for a high density project such as this. The connectivity between the various components of the master-plan on various levels was critical; the tower is integrated with the podium on various levels while retail and public uses at lower levels are integrated with the Metro system; the residential blocks are linked at the higher levels to create easier neighborhood accessibility while direct office and hotel connections are also provided for easier movement of people. The tower serves as a "Mini-city" which provides an amenity-rich focal point back to the community, offering a 24-hour city-life to be better for the environment and human interaction.

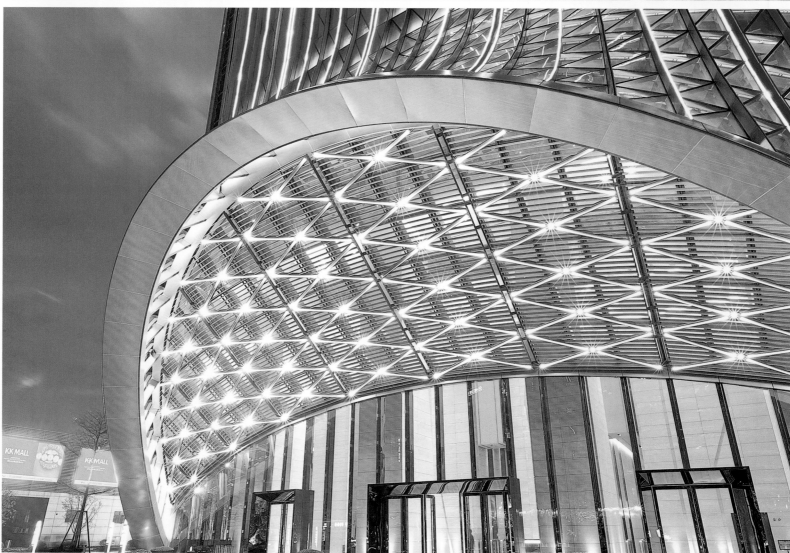

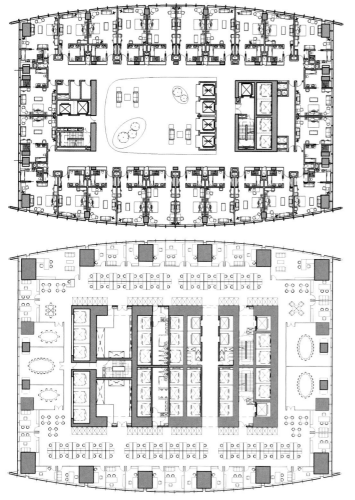

The public domain extends up the tower, with a "sky garden" housing restaurants and bars, as well as the hotel lobby, located right at the top of the building. This means that the public are not excluded from full enjoyment of the tower, which is all too often the case. The vertical circulation is user-friendly for operations as well public use: (1) Lifts for the office are kept very direct and simple with only one double height transfer lobby. This lobby again doubles up as a public space with opportunities for library, cafes, and viewing areas. (2) The hotel shuttle lifts brings the guests directly from ground-floor up to the 94th floor for check-in. The hotel lobby gives a unique experience and four local hotel lifts take visitors down to their rooms. (3) In case of fire, shuttle lifts will be used to assist total building evacuation.

One of the design features is the curving building profile. This form alludes to a spring or fountain and is intended to connote the wealth and prosperity of Shenzhen. The perimeter column arrangement provides each level with an unobstructed working environment and stunning views towards Lizhi and Renmin Park as well as over all Shenzhen and beyond.

It does not use the typical square foot print; the East / West façades being more slender and flared slightly so office floor plates are slightly bigger and the South / North façades that face Hong Kong and the Maipo marshes are wider. The slenderness brings certain challenges, most notably the swing or drift ratio and the robustness of the tower and performance of key elements. Instead of putting generators on top of the building, the roof is constituted by a curved smooth glazed curtain wall and steel structure.

With Shenzhen's growing population, clogged transport systems and an acute shortage of affordable land for development, the increased population density has become a major issue, therefore the logical key to a sustainable future is to build upwards. KK100, a major sustainable form of densification, will play an important contribution to meeting the ever-increasing demand for quality working and living space in the city.

The development accommodating large numbers of people into such a small footprint is better for the environment, as it puts less pressure on green spaces and local transport infrastructure while reducing suburban sprawl. This major mixed-use development promotes the idea of living and working in the same place, and reduces the need for commuting. This lessens reliance on the car, a polluting force, and eases the pressure on public transport. KK100 is a town centre in its own right.

From an environmental point-of-view, the advantage of facing primarily North-South (particularly in China) is a reduction in the East and West "heat gain" elevations. The vertical fins help to reduce low-level glare and provide shading. Also, they are important for the fixing of maintenance systems. Major "green" proposals included an environmentally friendly built form and envelope design; energy-saving building-services systems; a free-cooling system; and advanced building energy and environmental simulations. During the building's lifetime, the net aggregate of all these systems will contribute to the limitation of energy use and enhance the profile of the development as an environmentally aware and responsible contribution to the skyline of Shenzhen.

Plan 1 平面图 1

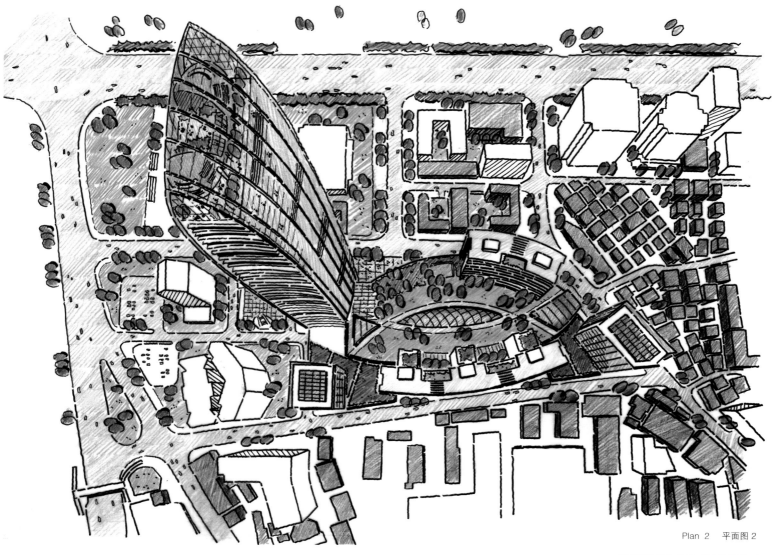

Plan 2 平面图 2

经世界高层都市建筑学会 (CTBUH) 认证，京基 100 是 2011 年度世界最高的建筑，这是一个具有创新性的高密度项目，为城市构成提供了全新的手法。它坐落在深圳市 CBD 区域的边界，为 21 世纪商业区域变革开创了先例。

建筑共 100 层，高 441.8m，总体规划 417 000 平方米的建筑综合体中，有 210 000 平方米被用作居住空间。该项目包括 5 栋住宅建筑和 2 栋商业建筑，总共划分为三个功能区域。4 至 72 层是甲级写字楼，面积 173 000 平方米；75 至 100 层是 35 000 平方米的 6 层级豪华酒店，配有一个如教堂般的玻璃顶的空中花园，可用来举办各种活动。

裙楼将这些元素联系到一起，并驱动着商业环境的发展，强调当地特色、刺激点和经济活力。这里将成为深圳最为卓越的商业活动区，而这也是它自身的发展目标。设计为商业区提供了多方位无死角的通道。

商业中心的停车场位于裙楼的三、四层，这意味着顾客从汽车出来后可以径直走到同层的购物中心。与传统的地下车库相比，这大大增强了车辆通道的便捷度。这样便利的设计更易于行人、业主及小区居民从各个方向进入购物中心，这使购物中心更像一个自然连接的主要街道，而不是一个封闭式的、独立的购物中心。实际上，购物中心扮演着连接器的角色，将与之发展相关的各个因素联系起来，同时很好地将其自身融入整条街道。

同时，京基 100 保证了社会与文化的连续性，与城市交通网络相结合，这对于一个高密度的工程来说是至关重要的。总体规划的各级各组成部分间的联系也是至关重要的；塔楼与裙楼在不同的楼层相结合，低层的商业零售区域和公共空间与地铁系统相结合；住宅在更高楼层处相联系，使邻里之间更加亲近，而办公空间与酒店的连接则为人们的穿行提供了便捷条件。塔楼就像是一个"微型城市"，服务于整个规划，为社区提供丰富的生活设施，实现 24 小时的城市生活以及人与环境之间的互动。

公共区域的部分使建筑得到延伸，包括附有"空中花园"的餐厅和酒吧，同时也包括位于建筑顶端的酒店大堂。这意味着人们不会像往常一样，无法在建筑中得

到充分的体验。垂直的交通组织对于管理者和公众来说都非常便捷：（1）办公区部分的电梯设计简洁明了，只有一个两层高的转换大厅，这个大厅作为公共空间也被分成了两个部分，包含图书室、咖啡厅和观赏区；（2）酒店部分的穿梭电梯可以运载着客人直接从地面层上升至 94 层，以办理入住手续，酒店大堂带给客人非同一般的感受，同时有 4 部酒店专用电梯可以运载客人下达至客房楼层；（3）一旦发生火灾，穿梭电梯亦可协助整座楼大楼的疏散。

本案的一个设计特点是曲线形的建筑轮廓。这种泉水式的流线造型喻示了深圳的财富与兴旺。建筑的网状排列为每一级空间都提供了畅通无阻的工作环境，同时可以享受荔枝公园与人民公园无与伦比的美景。

本案并没有采用通常的方形规划；而是将东西立面变得更加纤细，向外稍作倾斜，如此一来办公空间就会稍微扩展，同时使面向香港和米埔湿地的南北立面更加宽阔。这种纤细的规格给设计上带来了一定的挑战，最显著的就是摆动或层间位移角，结构的稳定性，还有重要部件的性能。设计师并未把发电机置于建筑的顶层，而是在顶部设计了光滑的曲线形玻璃幕墙和钢架结构。

随着深圳的人口增长，交通系统日益拥堵，土地资源匮乏，日益增长的人口密度已经成为一个重要问题，因此未来建筑必须向上建造。京基 100，将以一种可持续的形式，满足城市对工作品质和生活空间的不断增长需求。

将大量人口容纳到较小的建筑面积内，对环保来说极为有益的，在减小郊区扩张的同时也减小了生态空间和地方交通公共设施的压力。这种混合型功能建筑的开发推进了商住两用的概念，同时减少了对公共交通工具的需求，降低了对汽车的依赖，减少了污染，同时也缓解了公共交通的压力。京基一百本身就是市中心。 从环境角度来看，南北朝向的优势（特别是在中国）在于减少了东、西两面的"吸热"。垂直的鳍形结构减少了低层眩光并起到遮阳的作用，同样，它们对维护系统也起到了固定的重要作用。主要的"生态"提案包含一个环保建筑模式和基本框架设计；节能建筑服务系统；直冷系统；先进的建筑耗能和环境模拟系统。在建筑的整个使用期间，这些系统会利用有限的能源提高环保效用，并为深圳的天际增添一抹光彩。

HEADQUARTERS OF LHI PULLACH
普拉克 LHI 总部

Architects: Mann + Partner Freie Architekten BDA ,
Munich(A member of the SBA Group)
Client: LHI Campus Pullach GmbH + Co.KG
Location: Pullach, Munich, Bavaria, Germany
Building Area: 19,809 m²
Photographer: Martin Duckek, Christian Hacker ,
Mann + Partner Freie Architekten BDA

设计机构：Mann + Partner Freie Architekten BDA ,
Munich（SBA 集团成员）
客户：LHI Campus Pullach GmbH + Co.KG
项目地址：德国巴伐利亚慕尼黑市普拉克
建筑面积：19 809 平方米
摄影：Martin Duckek, Christian Hacker,
（Mann + Partner Freie Architekten BDA）

The Headquarters LHI is located in an idyllic hillside area near the Isar River. The design concept is to create a character which reminds of a campus feeling.

All facilities such as conference centre, canteen and office areas, including communication places and green spaces are developed as a combined space. For the scheme, tradition, identification and emotion are essential qualities. The building will be lowered by one floor, so that the public space will be not perceived. A pleasant working atmosphere will be created for about 350 employees. The variable system partitions are glazed generously. Thus, the central area receives abundant natural light and the working areas are linked with the landscape.

The staff canteen with a ceiling height of 6.50 meters also reflects these qualities. There are variable separations between the hall and the two-storey cafeteria. With the "Silent Gliss Electric Curtain System", the cafeteria can be separated from the atrium or gives a free view into the landscape behind the building. If the curtain is closed, there will be a harmonious and intimate atmosphere, supported by the inner courtyard.

The building ensemble opens down to the banks of the Isar. The landscape and the building are linked by the graded terrace levels.The green area descends from the south between the parts of structure by steel bands to the garden level.

The design of the natural courtyard provides the employees an enjoyable working place. By a water basin and a sliding green ramp, the environment enhances the well-being.

The energy concept acts without fossil energies. Therefore heating and cooling are widely covered by renewable energies.

LHI 总部位于伊萨尔河附近田园般的山坡上。其设计理念是创建一种校园般的感觉。

所有的设施如会议中心、餐厅和办公区，都是包括交流空间和绿色空间的组合空间。这一方案中，传统、可识别性和情感都是必备的品质。建筑将会隐藏一层，所以公共区域将不会被发现。

该项目将为约350名员工创造愉快的工作环境。可变的系统分区大量使用玻璃；因此，中央区域享有充足的自然光，工作区域和景观的联系更加紧密。

净高6.50米的员工食堂也可反映上述品质。大厅和两层的自助餐厅可以随意分隔。自助餐厅和中庭可利用电动窗帘系统分隔，可以享受建筑后面的景观。如果拉上窗帘，则可以在这里享受和谐而亲密的氛围。

建筑整体向伊萨尔河的河岸开放。景观和建筑通过梯级台地联系。绿色区域从南边通过钢铁带降至花园层。

自然庭院为员工们提供了愉快的工作场所。水池和变化的绿色坡道更强化了这种幸福感。

该项目的能源理念是不使用化石能源，因此加热和制冷均使用可再生能源。

Masterplan 规划图

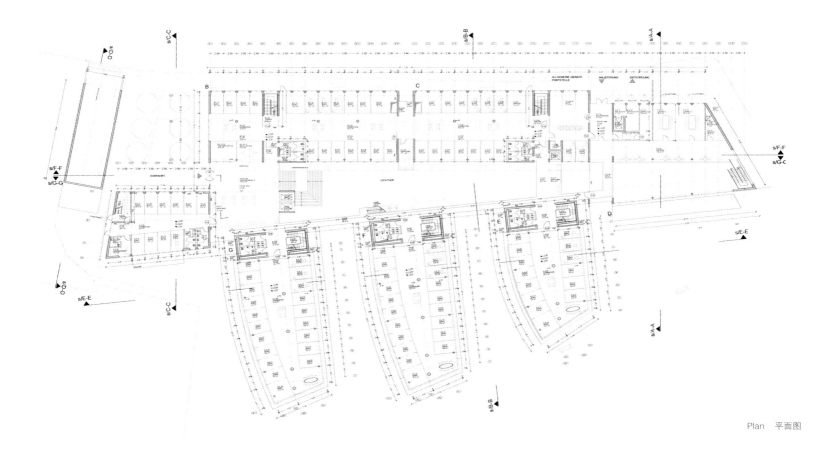

Plan 平面图

Profile Analysis 1 側面分析图 1

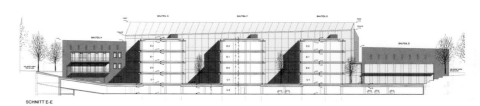

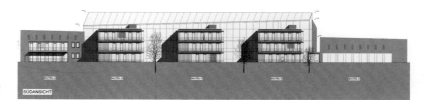

Profile Analysis 2 側面分析图 2

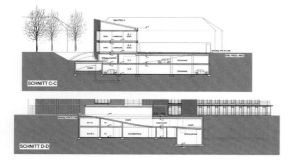

Profile Analysis 3 側面分析图 3

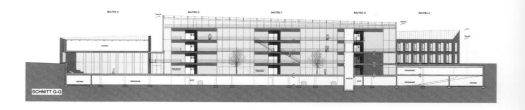

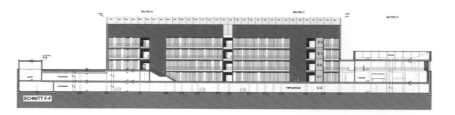

Profile Analysis 4 側面分析图 4

V TOWER
V 大厦

Architects: Wiel Arets
Client: Dis Valq, Valid
Project Team: Wiel Arets, Daniel Meier, Natali Gagro, Alex Kunnen
Urban Planner: MVRDV
Location: Eindhoven, the Netherlands
Area: 345,155 m²
Photographer: Jan Bitter, Christian Richters

设计机构：Wiel Arets
客户：Dis Valq, Valid
项目团队：Wiel Arets, Daniel Meier, Natali Gagro, Alex Kunnen
城市规划：MVRDV
项目地址：荷兰艾恩德霍芬
面积：345 155 平方米
摄影：Jan Bitter, Christian Richters

The V Tower is part of the Flight Forum in Eindhoven, an industrial park with an urban plan by MVRDV.
It is based on six clusters dropped loosely into the landscape and accessed from a looping road, a branch connection of the A2. The tower, which contains offices, is situated in"cluster 1", where all buildings with office functions are concentrated. The site is cut across its full length by a parking garage whose roof creates a platform for a second urban level, which even contains a bus line. While the main entrance of the tower is related to the elevated level, delivery and parking take place at the level of the ring road.
The thirteen-stacked floors correspond exactly with the geometry of the given polygonal footprint, resulting in a somewhat slender volume. Apart from the shared facilities, the entrance lobby and the lounge on the top levels, there is no difference in height between floors. The plan is programmed around two shafts that leave a bay's width to the exterior skin, thus allowing continuous circulation inside the facade. The inner organization of the shafts allows for a layout of cellular, as well as landscape offices. Besides the shafts, only three columns are needed to support the building.
The glazed outer surface is treated with an enamel print, effecting an alternation of opacity in the tower's skin. The transparency of the facade ranges from clear to diffuse in a seemingly random pattern, and depending on the light, density and angle, allows a glance inside the building or a reflection of the building's immediate surroundings.

　　V 大厦位于荷兰艾恩德霍芬 MVRDV 规划的一座工业园区 "Flight Forum" 里。

　　在这个规划中有 6 个组织松散的楼群，通过一条环路和 A2 公路的一条支路相连。V 大厦位于"楼群一"当中，那里所有的建筑都是办公楼。一个停车场横穿地块，其顶部设成平台，并设置了一条公交车线。大厦的主要入口位置较高，上下客和停车区域则在环路那层。

　　13 层的大厦采用了多边形的几何形状，看上去很"苗条"。除了公共设施、入口大堂和上部的休息厅之外，其他楼层的高度都是一样的。设计规划围绕两个轴展开，在其与外立面之间留出一定距离，便于人流通行。轴的内部组织考虑到单元与景观办公区的布局。除了这两个轴，这里仅需要 3 根圆柱支撑建筑。

　　玻璃外墙有着反光的银色装饰，在透明和反光之间不断变化。反射的效果取决于灯光、密度和角度。这样的外墙既能反射周围的景观，也能让人们透过它一睹其中的状况。

Plan 平面图

商业建筑 COMMERCIAL BUILDING

In the narrow sense, commercial buildings refer to public buildings used for exchange and circulation of commodities, as well as places where people are engaged in various business activities. Commercial buildings can be divided into retail stores where all kinds of daily necessities and means of production are sold, shopping malls and wholesale markets, trading places for financial securities, various service buildings such as hotels, restaurants, cultural entertainment facilities, shopping centers, commercial streets and clubs.

Modern commercial buildings emphasize not only the pursuit of commercial interest, but also social value, public interest, cultural taste and the impact on people s lifestyles. oo estimate whether a commercial building is provided with aesthetic value, we not only see its external factors such as decoration materials, facade design, proportion and color, but also emphasize the participation sense of people. Modern society is a people–oriented society, and people are the main body of the society and space. oo place people in an architectural building, the scale of the building should be considered carefully. Commercial buildings, especially large commercial building complex, should be complicated but not messy in form, bulky but not distorted, dignified but approachable, and its facade is also a symbol of commercial culture. ohe commercialization of architecture promotes the diversification of design approach in commercial buildings. After architects   endless uuests for innovation, current architectural design methods are gradually maturing. Plenty of new building terms are generated to enrich people s lives and meet people s physical and spiritual demands.

Modern commercial buildings play an important role in urban public buildings, impacting the development ,construction, image and environment of cities. ohey also play an important role in modern landscape, containing rich contents in people s daily lives, enriching the splendor of modern buildings through their unconventional, stylish and beautiful exterior designs. ohus modern commercial buildings have become the iconic symbol of modern civilization.

　　狭义上的商业建筑是指用来进行商品交换和商品流通的公共建筑，以及供人们从事各类经营活动的建筑物。商业建筑的主要类型有销售各类日常用品和生产资料等的零售店、商场和批发市场，金融证券等行业的交易场所，各类服务业建筑，如旅馆、餐馆、文化娱乐设施、购物中心、商业街和会所等。

　　现代商业建筑在追求商业利益的同时，也非常重视其自身的社会价值、公共利益、文化品位以及对人们生活模式带来的影响。判断一座商业建筑是否具有审美价值，不仅仅是看它的装修材料、立面设计、比例、色彩等外部因素，更重要的是强调人的参与意识。现代社会是一个以人为本的社会，人是社会的主体，也是空间的主体。建筑形体的尺度经过周密考虑，以使人能置身其中。商业建筑，尤其是大型商业综合体建筑的形体应该繁复而不凌乱、体量庞大而不失真的，它既是"超人"的又是"亲人"的，其外立面也是商业文化的表征。建筑的商业化促使了商业建筑外观设计手法的多样化，各式各样的设计五花八门、层出不穷。经过设计师们的努力创新，当下的建筑设计手法正在逐步走向成熟，并且产生了很多建筑类的新词汇，更好地丰富了人们的生活，满足了人们物质和精神的需要。

　　现代商业建筑是城市中的重要公共建筑，它对城市的开发和建设，对城市面貌和环境的塑造产生了重要影响，在现代景观中具有显著的地位，涵盖了人们日常所需的丰富内容。标新立异、时尚美观的外形设计使商业建筑为现代建筑增添了无限风采，成为现代文明的标志性符号。

KINGDOM TOWER
王国塔

Architects: Adrian Smith + Gordon Gill Architecture
Designer: Adrian Smith, Gordon Gill
Client: Jeddah Economic Company
Location: Jeddah, Saudi Arabia
Site Area: 530,000 m²
Height: 1,000 m
Photographer: Jeddah Economic Company,
Adrian Smith + Gordon Gill Architecture

设计机构：AS+ GG 建筑师事务所
设计师：Adrian Smith, Gordon Gill
客户：吉达经济公司
地址：沙特阿拉伯吉达
占地面积：530 000 平方米
高度：1 000 米
摄影：吉达经济公司，AS+ GG
建筑师事务所

At over 1,000 meters (3,280 feet) and a total construction area of 530,000 square meters (5.7 million square feet), Kingdom Tower will be the centerpiece and first construction phase of the $20 billion Kingdom City development in Jeddah, Saudi Arabia, near the Red Sea.

Expected to cost $1.2 billion to construct, Kingdom Tower will be a mixed-use building featuring a luxury hotel, office space, serviced apartments, luxury condominiums and the world's highest observatory. Kingdom Tower's height will be at least 173 meters (568 feet) taller than Burj Khalifa, which was designed by Adrian Smith while at Skidmore, Owings & Merrill.

AS+GG's design for Kingdom Tower is both highly technological and distinctly organic. With its slender, subtly asymmetrical massing, the tower evokes a bundle of leaves shooting up from the ground—a burst of new life that heralds more growth all around it. This symbolizes the tower as a catalyst for increased development around it.

The sleek, streamlined form of the tower can be interpreted as a reference to the folded fronds of young desert plant growth. The way the fronds sprout upward from the ground as a single form, then start separating from each other at the top, is an analogy of new growth fused with technology.

While the design is contextual to Saudi Arabia, it also represents an evolution and a refinement of an architectural continuum of skyscraper design. The three-petal footprint is ideal for residential units, and the tapering wings produce an aerodynamic shape that helps reduce structural loading due to wind vortex shedding. The Kingdom Tower design embraces its architectural pedigree, taking full advantage of the proven design strategies and technological strategies of its lineage, refining and advancing them to achieve new heights.

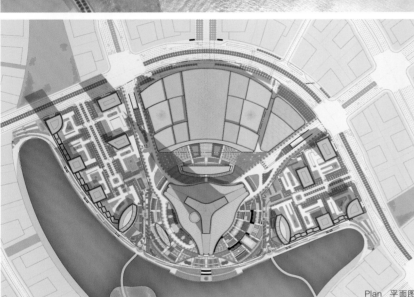

Plan 平面图

The result is an elegant, cost-efficient and highly constructible design that is at once grounded in built tradition and aggressively forward-looking, taking advantage of new and innovative thinking about technology, building materials, life-cycle considerations and energy conservation. For example, the project will feature a high-performance exterior wall system that will minimize energy consumption by reducing thermal loads. In addition, each of Kingdom Tower's three sides features a series of notches that create pockets of shadow that shield areas of the building from the sun and provide outdoor terraces with stunning views of Jeddah and the Red Sea.

The great height of Kingdom Tower necessitates one of the world's most sophisticated elevator systems. The Kingdom Tower complex will contain 59 elevators, including 54 single-deck and 5 double-deck elevators, along with 12 escalators. Elevators serving the observatory will travel at a rate of 10 meters per second in both directions. Another unique feature of the design is a sky terrace, roughly 30 meters (98 feet) in diameter, at level 157. It is an outdoor amenity space intended for use by the penthouse floor.

王国塔高 1 000 米（3 280 英尺）有余，用地面积 530 000 平方米（570 万平方英尺），它将成为造价 200 亿美元的王国城的一期工程项目的核心部分。该项目坐落在沙特阿拉伯吉达地区，临近红海。

王国塔预计耗资 12 亿美元，将成为一座多功能建筑，汇集豪华酒店、写字楼、酒店式公寓、豪华公寓以及全球最高的观光台。王国塔将比迪拜塔至少高出 173 米（568 英尺），而迪拜塔是 Adrian Smith 在 SOM 建筑设计事务所供职时的设计作品。

AS+GG 建筑师事务所设计的王国塔技术高超，自成一体。该塔高挑的塔身及其非对称的精巧外形，让人联想到一束叶片拔地而起——它代表了一种引领发展的全新生活，象征着王国塔将成为促进周边地区快速发展的催化剂。

该塔优美的流线型轮廓宛若生机勃勃的沙漠植物生长出的一簇闭合的叶片。"叶片"拔地而起，在顶部一分为二，这喻示着融合技术的新增长模式。

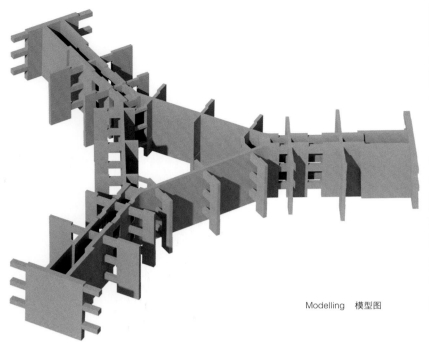

Modelling　模型图

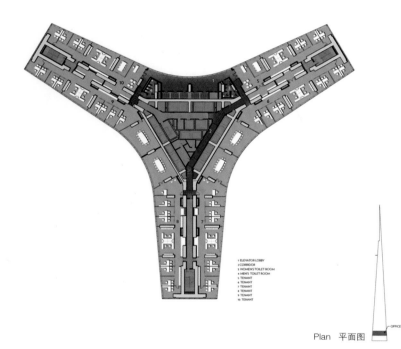

1 ELEVATOR LOBBY
2 CORRIDOR
3 WOMEN'S TOILET ROOM
4 MEN'S TOILET ROOM
5 TENANT
6 TENANT
7 TENANT
8 TENANT
9 TENANT
10 TENANT

OFFICE

Plan　平面图

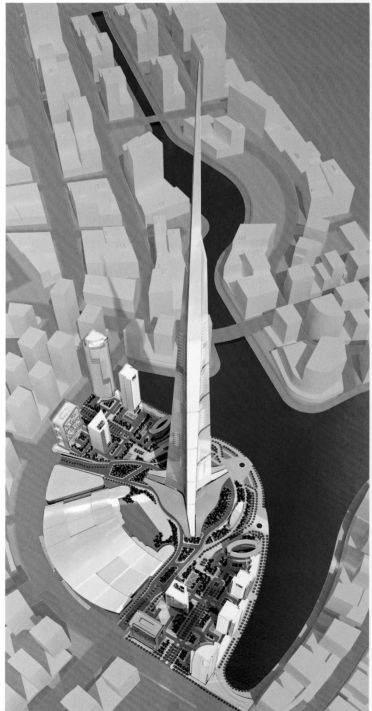

　　虽然该塔的设计与沙特阿拉伯的环境息息相关，但也代表了一种发展过程和对摩天大楼建筑综合体的改进。三个花瓣的区域是绝佳宜居之地，锥形侧翼形成了一种流线型外形，有助于降低因风涡旋而产生的结构荷载。王国塔的设计秉承了其建筑风格，充分利用了其世系的成熟设计策略和技术策略，并对这些策略进行改进，使其达到新的高度。

　　于是促成了一个优雅、投资效益好、可行性高的设计方案。该方案建立在固有的传统上，具有积极的前瞻性，并且运用了技术、建材、生命周期因素和节能等方面的创新思维。例如，该项目将采用高性能的外墙系统，通过减少热负荷而最大限度地减少能耗。此外，王国塔的三个侧面均设有一连串的缺口，可以产生口袋状的少许阴影，为建筑物的一些区域遮阳，并且构成了户外露台，将吉达与红海的壮丽景色尽收眼底。

　　王国塔的惊人高度使其必须具备世界上最复杂的电梯系统。王国塔建筑群将配备 59 部电梯，包括 54 部单轿厢和 5 部双轿厢电梯，以及 12 部自动扶梯。观光台的专用电梯将以双向每秒 10 米的速度运行。该设计的另一个独特功能是摩天台，直径约为 30 米（98 英尺），位于 157 层。这是供顶层复式楼使用的户外休闲空间。

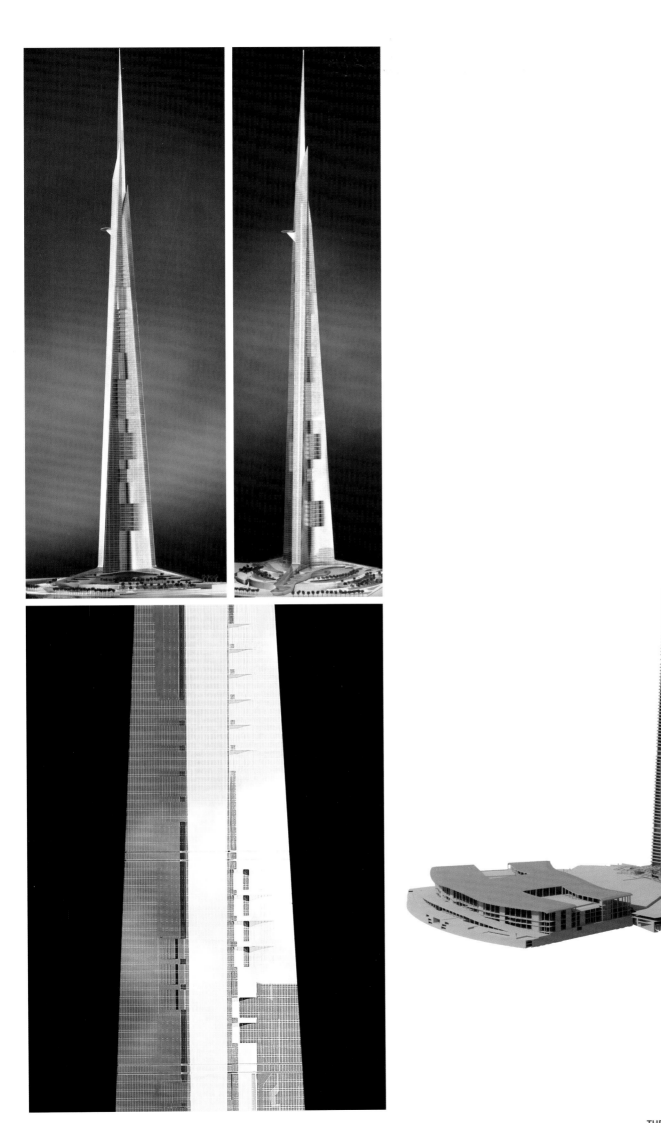

Modelling　模型图

DANCING DRAGONS
舞龙双子塔

Architects: Adrian Smith + Gordon Gill Architecture
Client: Yongsan International Business District, Seoul, Korea
Designer: Adrian Smith, Gordon Gill
Site Area: 23,000 m²
Location: Seoul, Korea
Photographer:Adrian Smith + Gordon Gill Architecture

设计机构：AS+ GG 建筑师事务所
客户：韩国首尔龙山国际商务区
设计师：Adrian Smith, Gordon Gill
占地面积：23 000 平方米
项目地址：韩国首尔
摄影：AS+ GG 建筑师事务所

Dancing Dragons is a pair of landmark supertall mixed-use towers for the new Yongsan International Business District in Seoul, Korea. The buildings, which include residential, official and retail elements, consist of slender, sharply angled minitowers cantilevered around a central core. The design aesthetic is highly contemporary yet informed by aspects of traditional Korean culture.

The minitowers feature a dramatic series of diagonal massing cuts that create living spaces that float beyond the structure. This recalls the eaves of traditional Korean pagodas — a design theme echoed both in the geometry of the building skin and the jutting canopies at the towers' base. The theme is extended in the building skin, which suggests the scales of fish and Korean mythical creatures such as dragons, which seem to dance around the core — hence the project's name. (Yongsan, the name of the overall development, means "Dragon Hill" in Korean.)

Dancing Dragons' scale-like skin is also a performative element. Gaps between its overlapping panels feature operable 600 mm vents through which air can circulate, making the skin "breathable" like that of certain animals.

Towers 1 and 2 — about 450 m and 390 m tall, respectively — share an architectural language and, therefore, a close family resemblance, but are not identical. In the taller structure, the 88-level Tower 1, the massing cuts at the top and bottom of the minitowers are V-shaped. In the 77-level Tower 2, the cuts move diagonally in a single unbroken line; they are also arranged in a radial pattern around the core that is perceptible as viewers move around the tower.

In both buildings, the minitower cuts are clad in glass at the top and bottom, making for dramatic skylights above the units at the highest levels and a transparent floor beneath the units at the lowest levels. This offers the opportunity for special high-value penthouse duplex units with spectacular 360-degree views of downtown Seoul and the adjacent Han River, along with an abundance of natural light.

The design of the 23,000m² site — part of the larger Yongsan master plan — reinforces the angular geometry of the buildings' massing and skin. Landscape features include sloped berms that echo that geometry. The site also includes a retail podium with a crystalline sculptural form and sunken garden that provides access to a large below-grade retail complex.

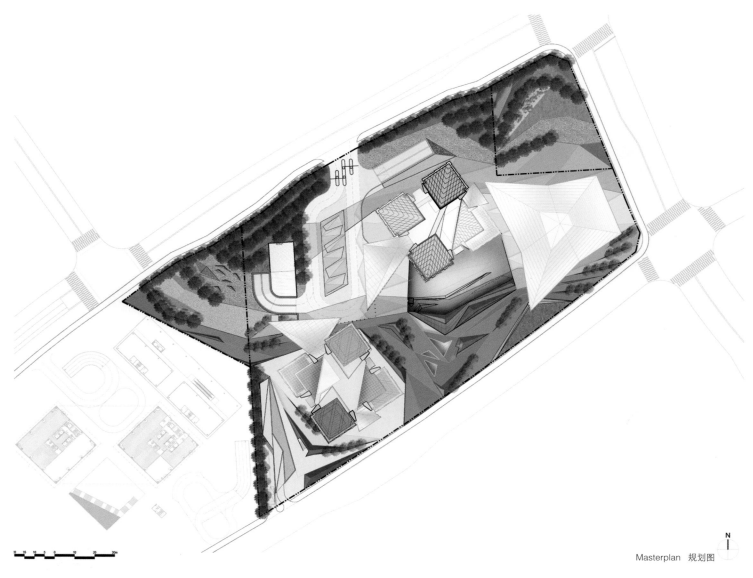

Masterplan 规划图

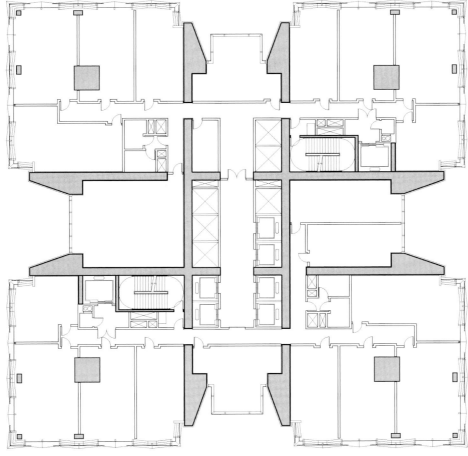

Plan 平面图

舞龙双子塔是韩国首尔龙山国际商务区两个超高层多用途的地标性建筑。这两座摩天大厦融住宅、办公与商业为一体，大量纤细陡峭的小塔楼悬吊在中央核心体上，其建筑美学颇具当代特色和韩国的传统文化特色。

所有小塔楼都有一系列的对角切口，创造出居住空间，悬浮于结构之上，让人联想到韩国传统宝塔的屋檐——这是建筑表皮和建筑基座上顶棚的设计主题。建筑表皮就像鱼鳞和巨龙的鳞片一样，使整个建筑仿佛围绕柱子舞动的巨龙，该建筑的名字便由此而来。（Yongsan 在韩语中就有"龙山"之意）

大厦鳞片般的表皮是建筑重要的特色。叠板之间的间隙有 600 毫米大的孔，可以使空气流通，使建筑像某种可自由呼吸的动物。

大厦 1 和大厦 2 分别高 450 米和 390 米，它们采用了相同的建筑语言，非常相似，但也有不同。88 层高的大厦 1 在顶部和底部建有小塔楼，它们都是"V"字形的。77 层高的大厦 2 的切口沿对角线移动，小塔楼沿核心按发射状排列。

每座大厦的小塔楼表面都覆盖有玻璃，在最高层形成独特的天窗，在最低层形成透明的地表。这可以使建筑里的单元拥有首尔和汉江 360 度的全景视野，并可享受充足的自然光。

这个 23 000 平方米的地块作为大龙山区总体规划的一部分，加强了建筑体量和表皮的几何学特征。倾斜的步道是景观方面的特色，这与建筑的几何特征相呼应。建筑还包括像水晶雕塑般的商业裙房和下沉式花园，花园有入口通往下层的零售区域。

Elevation　立面图

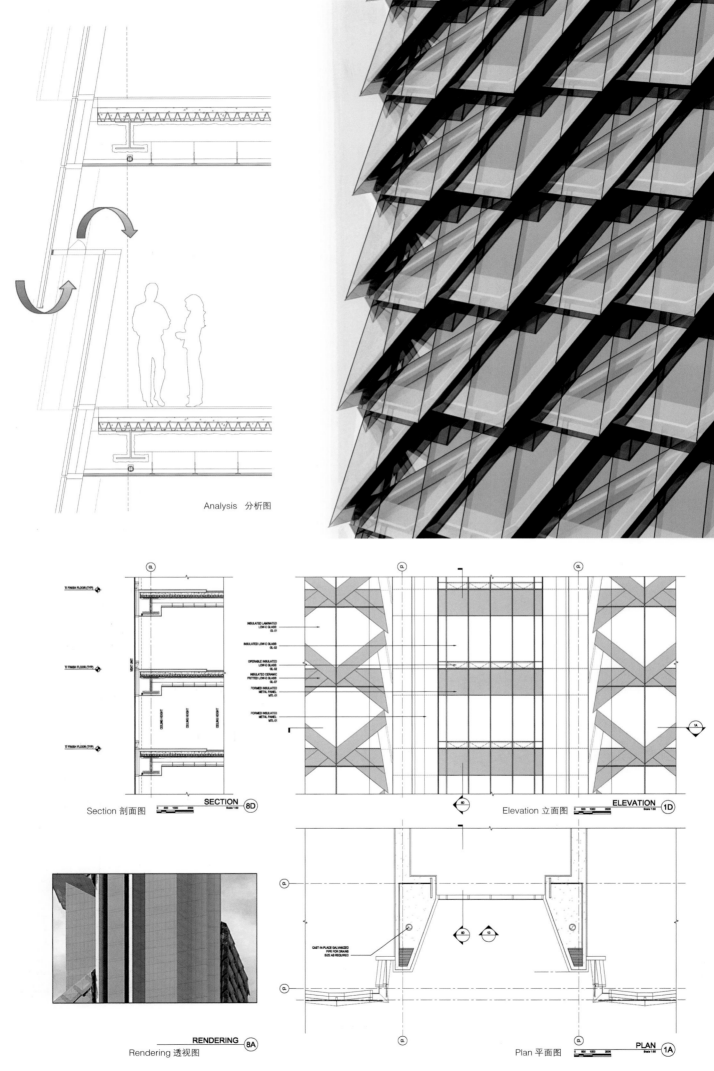

Analysis 分析图

Section 剖面图   SECTION 8D

INSULATED LAMINATED
LOW-E GLASS
GL-01

INSULATED LOW-E GLASS
GL-02

OPERABLE INSULATED
LOW-E GLASS
GL-02

INSULATED CERAMIC
FRITTED LOW-E GLASS
GL-07

FORMED INSULATED
METAL PANEL
MTL-01

FORMED INSULATED
METAL PANEL
MTL-01

Elevation 立面图   ELEVATION 1D

RENDERING 8A
Rendering 透视图

CAST-IN-PLACE GALVANIZED
PIPE FOR DRAINAGE
SIZE AS REQUIRED

Plan 平面图   PLAN 1A

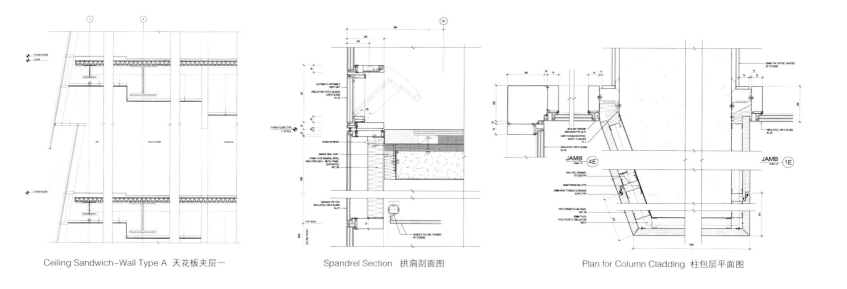

Ceiling Sandwich-Wall Type A 天花板夹层一   Spandrel Section 拱肩剖面图   Plan for Column Cladding 柱包层平面图

ZA'ABEEL ENERGY CITY MASTER PLAN
ZA'ABEEL 能源城市总体规划

Architects: Adrian Smith + Gordon Gill Architecture
Designer: Adrian Smith, Gordon Gill
Location: Dubai, United Arab Emirates
Photographer: Adrian Smith + Gordon Gill Architecture

设计机构：Adrian Smith + Gordon Gill Architecture
设计师：Adrian Smith, Gordon Gill
项目地址：阿拉伯联合酋长国迪拜
摄影：AS + GG 建筑师事务所

The Za'abeel Energy City Master Plan is uniquely positioned as a centre for both commercial and residential development. The project will embody modern, sustainable living, working and recreation in Dubai.

With connections by rail to Jumeira Gardens and convenient proximity to the Burj Dubai downtown, the project will be a vibrant mixed-use centre. The generative concept behind the Master Plan is the creation of memorable places that define sustainable districts and neighbourhoods.

The Master Plan proposes the creation of a major new civic park space. All residents and workers will have direct access to the commons through shaded pedestrian passageways. Multi-modal transportation access to the project and opportunities for pedestrian and bicycle use within the development provide all users with a diverse experience and due to its proximity to transit, the Master Plan has an opportunity to become a self-contained, mixed-use environment and achieve Dubai's first LEED platinum rating for community design.

Office and hotel functions add interest along the main boulevard and throughout the development, effectively reinforcing the connection to convention and hospitality functions of this new business centre. Private, luxury residential lofts define the perimeter of the project, providing a unique and intimate living experience.

In addition to creating memorable urban places, the Master Plan utilizes Islamic patterns and design traditions, re-interpreted for modern materials and lifestyles, to create variation and uniqueness in building designs and landscapes. These patterns, drawn from Islamic textiles, architectural ornament, and traditional village forms add texture and scale to the project.

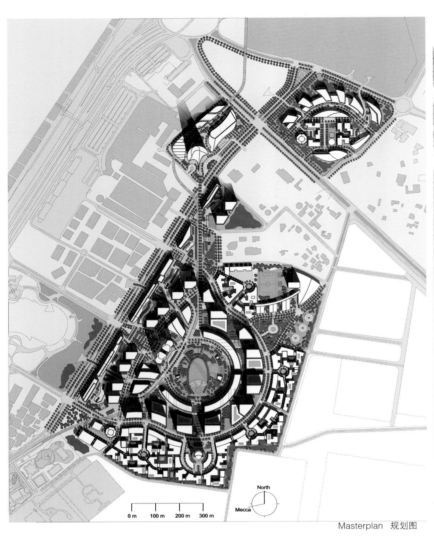

Masterplan 规划图

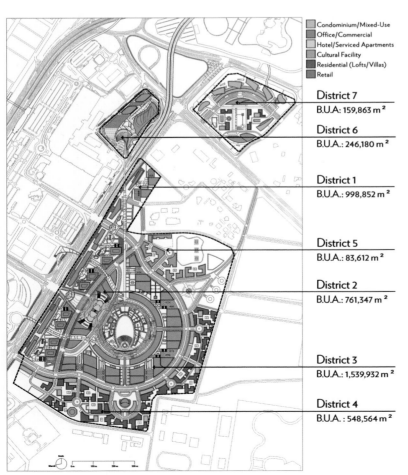

District 7
B.U.A.: 159,863 m²

District 6
B.U.A.: 246,180 m²

District 1
B.U.A.: 998,852 m²

District 5
B.U.A.: 83,612 m²

District 2
B.U.A.: 761,347 m²

District 3
B.U.A.: 1,539,932 m²

District 4
B.U.A.: 548,564 m²

Analysis 分析图

Za'abeel 能源城市总体规划项目定位为商业和住宅中心，体现迪拜的现代、可持续生活、工作和娱乐方式。

项目邻近迪拜市中心，与 Jumeira 花园由轨道相连，是一个充满生机的多功能中心。总体规划的设计概念是创造一个具有纪念意义的场所，界定出可持续发展的区域和邻区。

总体规划提出建设一个新的市民公园，所有的居民和工人通过人行通道可直接到达公园。规划区内有多样化的交通出入口可供行人和骑自行车者使用，这带给人们不同的体验。由于和运输线邻近，项目有可能成为自给自足的多功能发展区，成为迪拜第一个得到 LEED 认证的社区设计。

办公和酒店功能为林荫大道和整个发展区增加了趣味性，有效增强了这个商业中心的会议功能和接待功能。私密奢华的住宅顶楼界定了项目的边界，提供了一种独特而私密的生活体验。

除了创造具有纪念意义的城市空间，总体规划还采用伊斯兰设计风格，重新诠释了现代材料和现代的生活方式，创造出多变和独特的建筑设计和景观。这些生活方式来源于伊斯兰纺织品、建筑装饰和传统的乡村之中，这样提升了项目品质、扩大了项目规模。

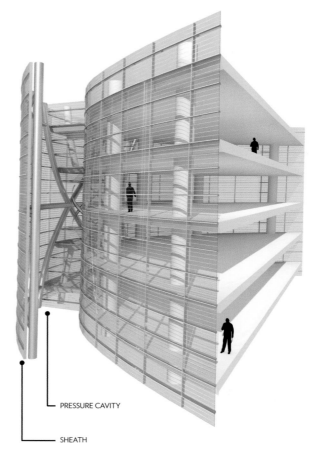

PRESSURE CAVITY

SHEATH

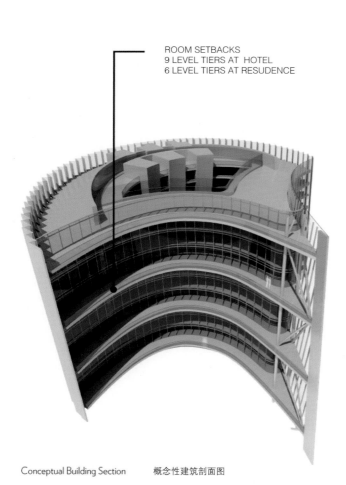

ROOM SETBACKS
9 LEVEL TIERS AT HOTEL
6 LEVEL TIERS AT RESUDENCE

Conceptual Building Section　概念性建筑剖面图

View of Photovoltaic Fin　光电散热视图

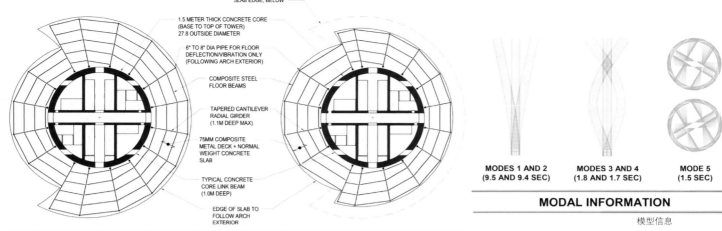

SLAB EDGE, BELOW

1.5 METER THICK CONCRETE CORE
(BASE TO TOP OF TOWER)
27.8 OUTSIDE DIAMETER

6" TO 8" DIA PIPE FOR FLOOR
DEFLECTION/VIBRATION ONLY
(FOLLOWING ARCH EXTERIOR)

COMPOSITE STEEL
FLOOR BEAMS

TAPERED CANTILEVER
RADIAL GIRDER
(1.1M DEEP MAX)

75MM COMPOSITE
METAL DECK + NORMAL
WEIGHT CONCRETE
SLAB

TYPICAL CONCRETE
CORE LINK BEAM
(1.0M DEEP)

EDGE OF SLAB TO
FOLLOW ARCH
EXTERIOR

MODES 1 AND 2
(9.5 AND 9.4 SEC)

MODES 3 AND 4
(1.8 AND 1.7 SEC)

MODE 5
(1.5 SEC)

MODAL INFORMATION
模型信息

TYPICAL FRAMING PLANS
典型构架平面图

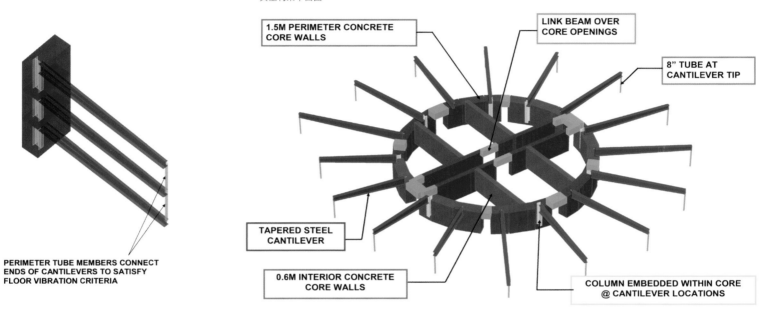

1.5M PERIMETER CONCRETE
CORE WALLS

LINK BEAM OVER
CORE OPENINGS

8" TUBE AT
CANTILEVER TIP

TAPERED STEEL
CANTILEVER

0.6M INTERIOR CONCRETE
CORE WALLS

COLUMN EMBEDDED WITHIN CORE
@ CANTILEVER LOCATIONS

PERIMETER TUBE MEMBERS CONNECT
ENDS OF CANTILEVERS TO SATISFY
FLOOR VIBRATION CRITERIA

CANTILEVER BEHAVIOR
悬臂操作

3D ANALYSIS MODEL: TYPICAL FLOOR
三维分析模型：标准层

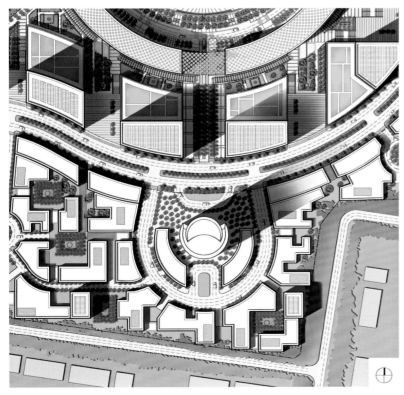

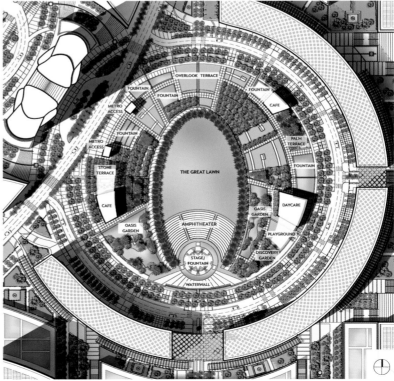

ME BARCELONA HOTEL
巴塞罗那 ME 酒店

Architects: Dominique Perrault Architecte
Client: Hoteles Sol-Melià, Palma de Mallorca, Spain
Associate Architects: Corada Figueras Arquitectos
Local Architects: AIA Salazar-Navarro
Project Manager: Tag Management
Site Area: 3,230 m²
Building Area: 29,334 m²
Height : 117 m
Location: Barcelona, Spain
Photographer: DPA/ADAGP

设计机构：多米尼克·佩罗建筑师事务所
客户：Hoteles Sol-Melià, Palma de Mallorca, Spain
合作机构：Corada Figueras Arquitectos
当地设计机构：AIA Salazar-Navarro
项目经理：Tag Management
占地面积：3 230 平方米
建筑面积：29 334 平方米
建筑高度：117 米
项目地址：西班牙巴塞罗那
摄影：DPA/ADAGP

Designed for the Habitat Group in Barcelona and now managed by ME, this hotel integrates the two dimensions that compose the identity of the Catalonian capital: the horizontality of its grid, legacy of the Cerdà plan, extending all the way to the sea, and its dynamic verticality exemplified by the Sagrada Familia and Mount Tibidabo looming over the sight.

The tower is composed of two volumes stuck together : a "cubic" building acting as a counterpoint and a tower 120 m high, a rectangular parallelepiped cut lengthwise in two. A cantilever, 20 m above street level, marks the entrance: on the Avinguda Diagonal it serves as the Hotel's identifying signal.

The way these boxes are placed against each other is the key to the distribution of the various functions. While the volume located at the back gathers the hotel's collective services, the tower, broad but not deep, houses the 259 guest rooms, each with a clear perspective of the scenery.

This "enormous screen " that focuses on the city and the landscape is cut into opaque panels of distinct texture that cover the entire facade, making it come alive in the day and the night.

The hotel is compound by 259 guest rooms (192 supreme rooms, 44 superior rooms, 16 suites, 6 grand suites, 1 sky suite, 4 double rooms with access for disabled people), a fitness centre, a restaurant (300 m²), a conference centre (1,150 m²), a swimming pool and terraces, a bar, salons, administration and underground car park.

Rendering 1　透视图 1

Rendering 2　透视图 2

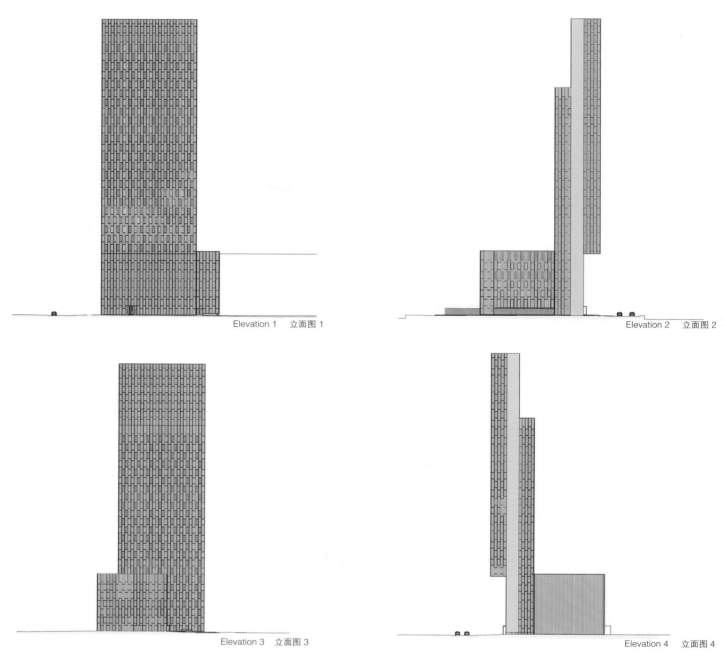

Elevation 1　立面图 1

Elevation 2　立面图 2

Elevation 3　立面图 3

Elevation 4　立面图 4

Plan　平面图

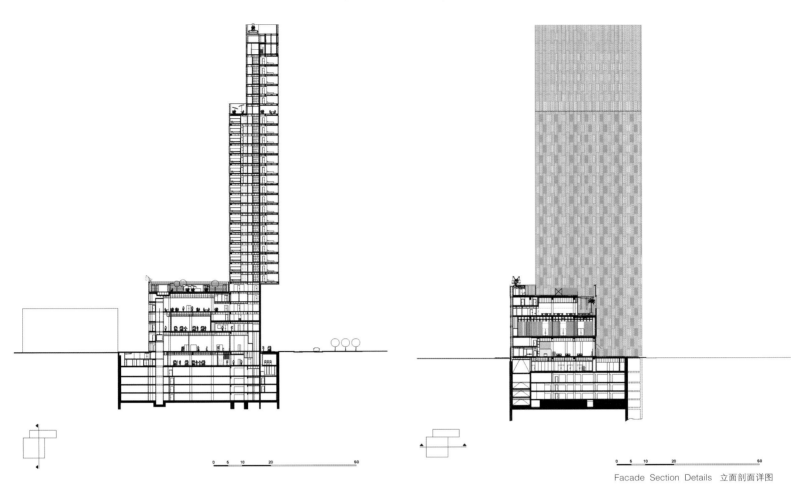

Facade Section Details 立面剖面详图

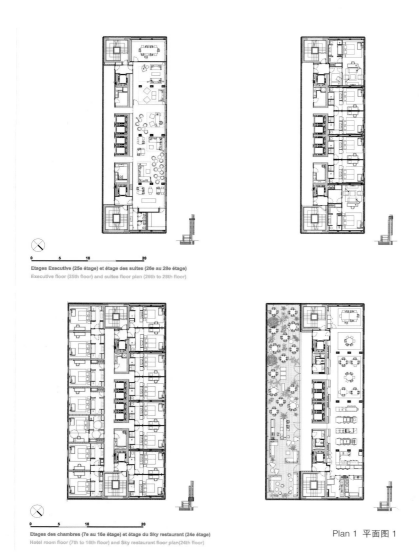

Etages Executive (25e étage) et étage des suites (26e au 28e étage)
Executive floor (25th floor) and suites floor plan (26th to 29th floor)

Etages des chambres (7e au 16e étage) et étage du Sky restaurant (24e étage)
Hotel room floor (7th to 16th floor) and Sky restaurant floor plan(24th floor)

Plan 1 平面图 1

这家酒店原先为巴塞罗那的 Habitat 集团设计,如今由 ME 接管。酒店集加泰罗尼亚首府——巴塞罗那城市特色的两大方面于一身:一是城市的水平网格,这是塞尔达城市规划的遗迹,所有道路都通向大海;另一个是动态的垂直风景,笼罩视野的圣家堂和蒂维达沃山就是最佳实例。

这座大楼由两个体块"粘"在一起:一座立方体建筑作为基座,一座 12 米高的长方体塔楼,沿纵向被一分为二。一个悬臂结构距地面 20 米高,指示了入口位置:在 Avinguda Diagonal 大道上,让人一眼就能认出酒店所在的位置。

这些盒子状的体块互相叠放的方式是分布各种功能的关键。位于后方的体块集中了酒店的各种服务功能,进深不大却宽敞的塔楼容纳了 259 间客房,每个房间都能清晰地欣赏到窗外美景。

这个"巨大的屏幕"聚焦于城市和景观,被切割成纹理截然不同的不透明面板,覆盖着整个立面,无论白天黑夜,酒店立面看起来都活力四射。

酒店由 259 间客房(192 间尊贵套房,44 间高级客房,16 间套房,6 间豪华套房,1 间天空套房,4 间双人房设有残障人士通道)、一个健身房、一个餐厅(300 平方米)、一个会议中心(1 150 平方米)、一个游泳池和露台、一个酒吧以及沙龙、行政管理区、地下停车场组成。

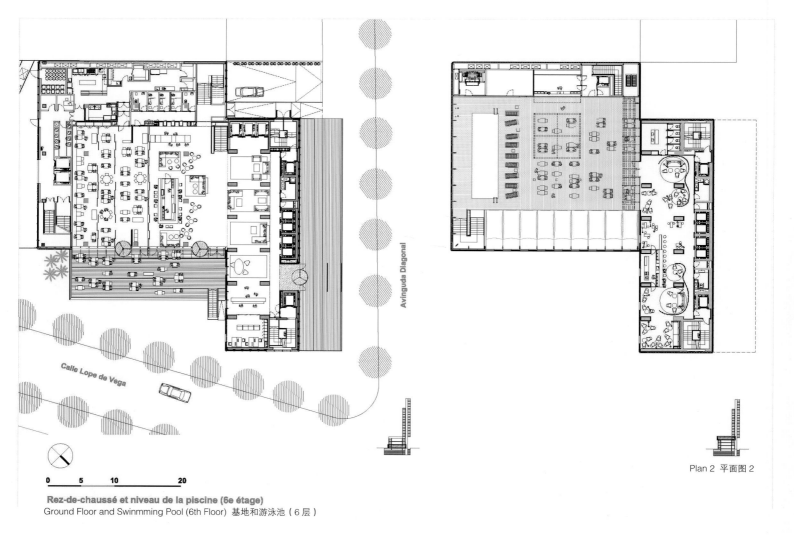

Rez-de-chaussé et niveau de la piscine (6e étage)
Ground Floor and Swinmming Pool (6th Floor) 基地和游泳池(6层)

Plan 2 平面图 2

CITY OF DREAMS
新濠天地

Architects: ARQUITECTONICA
Designer: Bernardo Fort-Brescia
Location: Macao, China
Area: 381,600 m²
Photographer: Photo Courtesy City of Dreams

设计机构：ARQUITECTONICA
设计师：Bernardo Fort-Brescia
项目地址：中国澳门
面积：381 600 平方米
摄影：Photo Courtesy City of Dreams

Located in the northeast portion of the Cotai strip, next to Macao University of Science and Technology, City of Dreams sets many facilities for entertainment, night clubs, hotels, restaurants, shops and casinos.

Four hotel towers rise from a casino podium that defines a water theme for the development. Each hotel tower establishes its identity while participating in the overall architectural message. Two towers comprise the Grand Hyatt Hotel. Located at the eastern side of the podium, the buildings rise in the form of two slender slab towers. The Crown Hotel stands in the northwestern corner of the site. This elliptical tower rises from the expansive series of reflecting pools and water features that define the main frontage of the project along the Cotai strip.

The Hard Rock Hotel is iconic in its own way, playful and young. Its pure circle provides efficiency and reception of room modules. It appears as a glass cylinder with a series of curved octagons cantilevering beyond its skin. These bands are rotated gradually to create a spiraling effect by the movement of shadows created by the horizontal eyebrows. The added movement to the cylinder is almost like a water spout or vertically flowing water cascade. They form a three dimensional sculpture that moves both vertically and horizontally carrying the eye around the form. The result is simple and complex, fun and interesting, architectural and theatrical.

The ends of the first Grand Hyatt tower curve as they rise, creating a vertical flowing movement. The slab form of the tower shifts in plan along its long axis emphasizing the slenderness and verticality of this building and enhancing the sense of movement. An array of horizontal fins projecting from the building skin undulates across the facade following the form of the curved ends. The second Grand Hyatt tower compliments the composition. A similar treatment to the slab form is used in this tower however the implied movement is horizontal. The roof parapet and main lobby soffit profiles curve across the length of the building in creating a wave like form. Vertical fins are arrayed across the facade following the parapet and soffit profiles again emphasizing the movement of the form.

The pure curved facade of the building is interrupted by a series of long vertical fins, each breaking up as they descend, creating a cascade effect as the facade disappears into the water at its base. A similar glass type is used on all towers, with their striking yet complimentary forms defining their identity.

Rendering 1　透视图 1

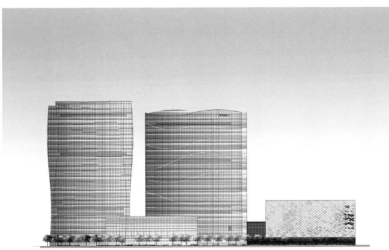

Rendering 2　透视图 2

Rendering 3　透视图 3

Rendering 4　效果图 4

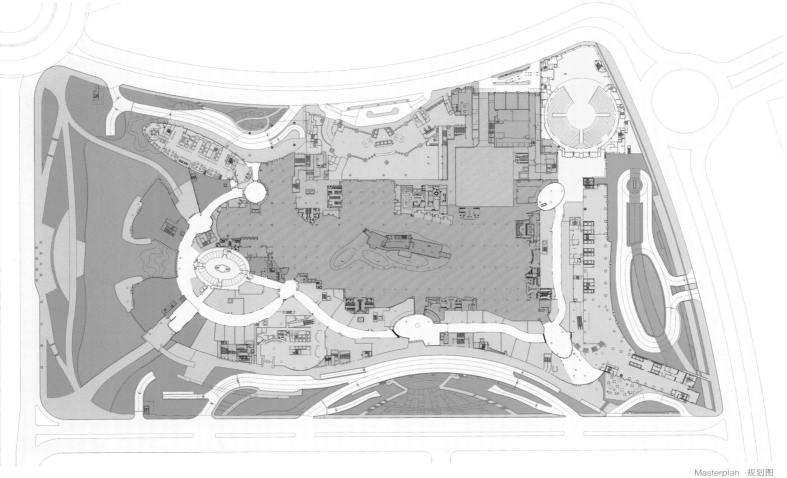

Masterplan 规划图

澳门新濠天地位于金光大道东北侧，毗邻澳门科技大学，集娱乐设施、夜店、酒店、餐厅、商场、赌场于一身。

四个酒店大厦围绕着以水为主题的娱乐场地拔地而起。每一个大厦在融入整体建筑特色的同时具有其独自的特征。君悦酒店由其中两个大厦组成。位于娱乐场地东侧的，是两个形态细长的大厦。皇冠度假酒店位于场地西北角。这个椭圆形建筑在一系列的倒影池和水景中升起。这也正是该项目面向金光大道立面的主要特征。

Hard Rock 酒店以其自身的趣味性和年轻态为特色。圆形结构为酒店提供了高效的房间接待能力。酒店造型仿佛一个玻璃圆筒，建筑表皮采用一系列八角形的悬臂结构。这些八角形逐层旋转，利用水平屋顶窗形成的阴影，创造出螺旋形的视觉效果，如旋风或瀑布一般。最终在人们的视线范围内形成一个能够进行垂直或水平运动的三维立体雕塑，简洁而复杂，颇具趣味性和戏剧性。

君悦酒店 A 座两端的曲线形设计创造了垂直的流体动感，别具匠心。大厦的塔层结构沿垂直轴分层，格外凸显了该建筑垂直细长及动感的特点。从建筑表面衍生的水平的鳍状结构沿立面起伏，形成了建筑的曲线形轮廓。君悦酒店 B 座的结构亦值得称道。类似 A 座的板塔形式在 B 座再现，但隐含的动感呈现水平态。屋顶栏杆和大堂拱腹构成曲面横穿建筑，以呈现波形的结构。垂直的鳍状结构横穿立面，栏杆和拱腹设计再次强调了动感的形态。

建筑物的弧形外观由一系列细长垂直的鳍状元素划分，营造出瀑布般的效果。一个类似玻璃的设计结构被用于所有塔座的设计之中，以其开放式的姿态诠释了它们的定位。

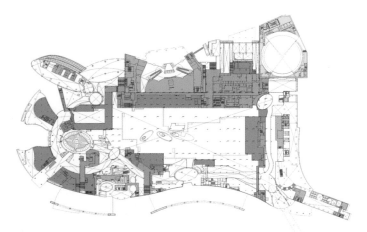

Masterplan 1 规划图 1

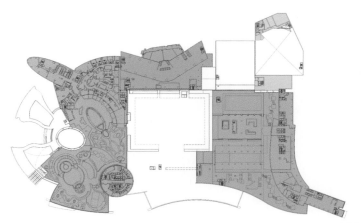

Masterplan 2 规划图 2

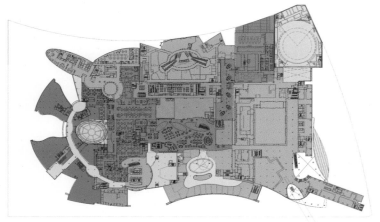

Masterplan 3 规划图 3

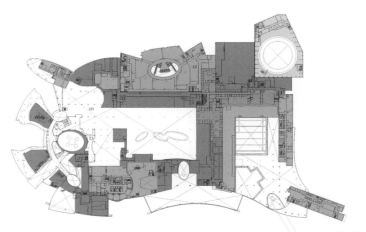

Masterplan 4 规划图 4

HORIZONTAL SKYSCRAPRE —VANKE CENTRE
卧式摩天楼 —— 万科中心

Architects: Garrick Ambrose (SD/DD), Maren Koehler (DD), Jay Siebenmorgen (DD), Christopher Brokaw (CD), Rodolfo Dias (CD)
Project Managers: Yimei Chan, Gong Dong
Design Architects: Steven Holl, Li Hu
Assistant Architect: Eric Li
Location: Shenzhen, Guangdong, China
Area: 120,445 m²

设计机构：Garrick Ambrose (SD/DD), Maren Koehler (DD), Jay Siebenmorgen (DD), Christopher Brokaw (CD), Rodolfo Dias (CD)
项目经理：Yimei Chan, Gong Dong
设计建筑师：Steven Holl, Li Hu
助理建筑师：Eric Li
项目地址：中国广东深圳
面积：120 445 平方米

Project Team：Jason Anderson, Guanlan Cao, Clemence Eliard, Forrest Fulton, Nick Gelpi, M. Emran Hossain, Kelvin Jia, Seung Hyun Kang, JongSeo Lee, Wan–Jen Lin, Richard Liu, Jackie Luk, Chris McVoy, Enrique Moya–Angeler, Roberto Requejo, Michael Rusch, Jiangtao Shen, Filipe Taboada, Manta Weihermann

项目团队：Jason Anderson, Guanlan Cao, Clemence Eliard, Forrest Fulton, Nick Gelpi, M. Emran Hossain, Kelvin Jia, Seung Hyun Kang, JongSeo Lee, Wan–Jen Lin, Richard Liu, Jackie Luk, Chris McVoy, Enrique Moya–Angeler, Roberto Requejo, Michael Rusch, Jiangtao Shen, Filipe Taboada, Manta Weihermann

Hovering over a tropical garden, this "horizontal skyscraper" — as long as the Empire State Building is tall — is a hybrid building including apartments, a hotel, and offices for the headquarters for China Vanke Co. Ltd. A conference center, spa and parking are located under the large green, tropical landscape, which is characterized by mounds containing restaurants and a 500-seat auditorium.
The building appears as if it was once floating on a higher sea that has now subsided; leaving the structure propped up high on eight legs. The decision to float one large structure right under the 35-meter height limit, instead of several smaller structures each catering to a specific program, generates the largest possible green space open to the public on the ground level.

这座卧式摩天楼为一个建筑综合体，盘旋在热带公园上空，高耸入云，包括公寓、酒店以及万科中心总部办公区。会议中心、水疗馆以及停车场均位于大面积的绿色热带景观之中。景观以沙丘为特色，设有餐厅和可容纳 500 人的礼堂。
该建筑看上去就像漂浮在海平面上，平静且安稳，是由 8 根支柱支撑起来的。庞大的建筑在限高 35 米的条件下漂浮，而不是采取几个较小的结构以迎合某项特定工程。这一设计最大限度地形成了绿色空间，并完全向公众开放。

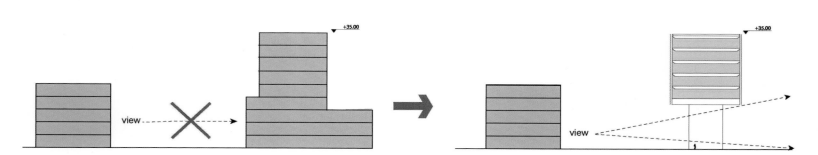

01. Development as SWA originally planned

02. Solution creates a mutually beneficial condition

Unobstructed Views 畅通无阻的视线

118

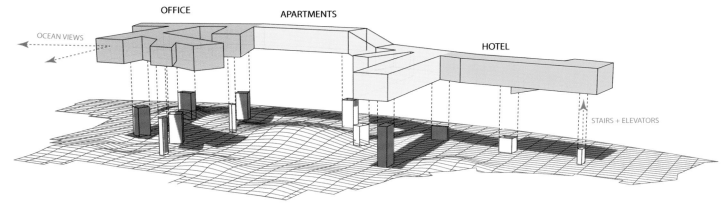

OFFICE APARTMENTS HOTEL

OCEAN VIEWS

STAIRS + ELEVATORS

LANDSCAPE

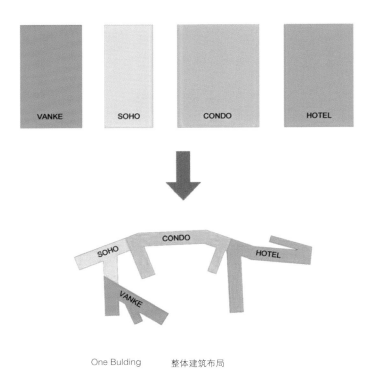

VANKE SOHO CONDO HOTEL

SOHO CONDO HOTEL

VANKE

One Bulding 整体建筑布局

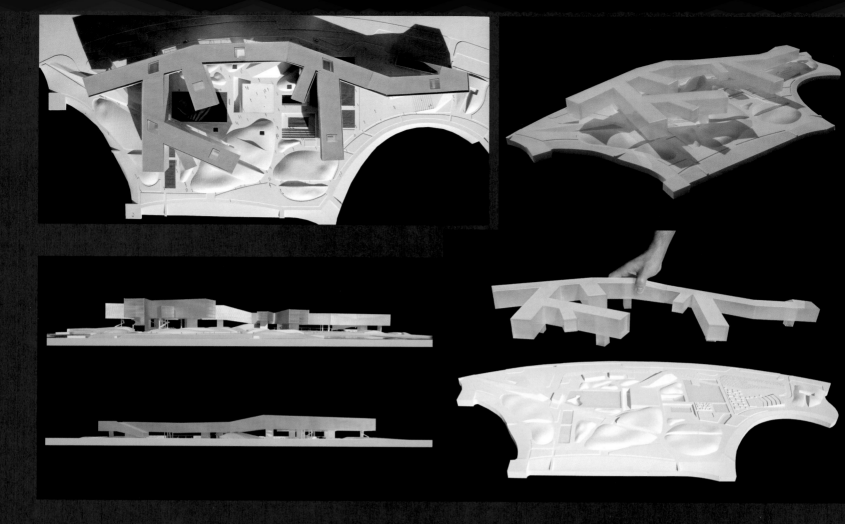

Modelling　模型图

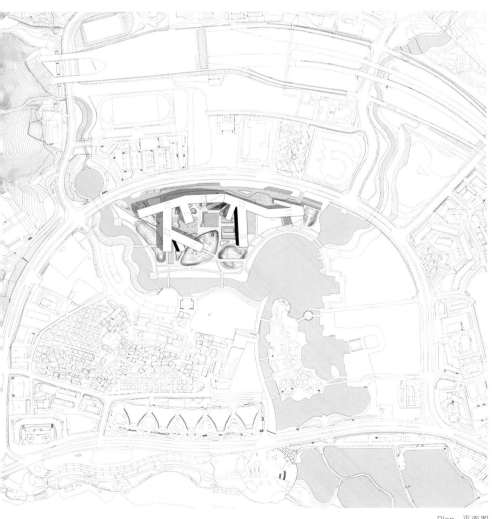

Plan 平面图

D-CUBE CITY
多客福城

Architects: The Jerde Partnership
Perform Designer: Samoo
Area: 320,000 m²
Location: Seoul, Korea

设计机构：捷得建筑师事务所
执行设计师：Samoo
面积：320 000 平方米
项目地址：韩国首尔

D-Cube City, located in Seoul, the dense capital city of Korea, sets a new standard in mixed-use transit-oriented development connected to the city's busiest metro line. The new cultural and commercial destination is one of the city's first fully integrated developments of its kind, made up of over 300,000 m² of high-rise office and hotel, a multi-level commercial retail, entertainment and cultural complex, and over 24,281 m² of public landscape, parks and plazas.
Immediately adjacent and connected to the Shindorim Station, the project creates a global example of sustainable, transit-oriented development resulting in urban regeneration and social advancement. The innovative transformation of the site into a mixed-use public district represents a major accomplishment for land redevelopment in Korea and is expected to be a catalyst for the continued growth and evolution of the area into a vibrant urban hub.

多客福城坐落于人口密集的韩国首都——首尔，它与城市最繁忙的地铁线相连，为多用途公共交通导向式发展设立了新标准。这个新的文化商业中心是整座城市同类建筑中第一批全面综合发展的范例之一，它由建筑面积超过 300 000 平方米的各类建筑组成：包括高层写字楼和酒店、一个多层商业零售中心、娱乐文化设施以及超过 24 281 平方米的公共景观、公园和广场。
项目建设区直接与邻近的新道林地铁站相连，创造了一个可持续的、公共交通导向式发展的全球典范，促进了城市更新和社会进步。项目场地创造性地转变为一个多功能公共区，代表了韩国在土地利用再发展中的一个重要成就，在持续刺激发展的同时，此地区将发展成一个充满活力的城市中心。

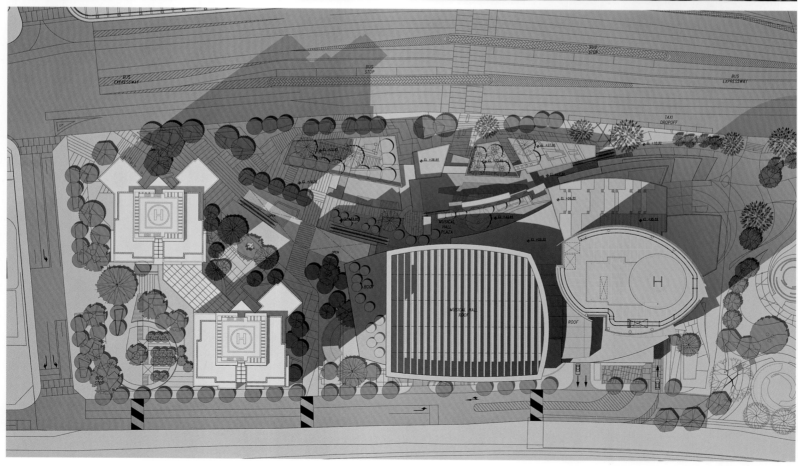

Masterplan 规划图

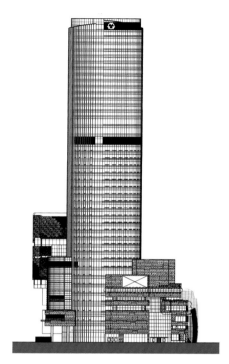

Elevation 1 立面图 1

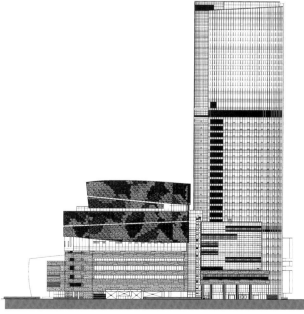

Elevation 2 立面图 2

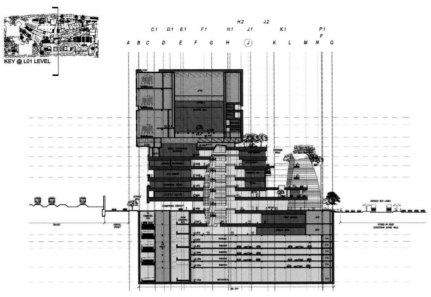

Analysis 1 分析图 1

Intended to formulate a co-existence of nature and culture within a highly dense urban environment, D-Cube City's artful vertical design incorporates reminiscent elements of traditional Korean landscape paintings of endless mountains and rivers. Among the project's design highlights are building forms organic to Korea that are shaped like Asian lanterns which create warm, glowing light filtering through the exterior cladding at night to draw visitors into the project. There is also an experiential outdoor pathway climbing through the lantern buildings to the top of the retail complex that has the character of an Italian hill town executed in modern contemporary architecture.

Unique to Seoul, D-Cube City's design interweaves natural expressions to create an urban oasis that redefines the district's industrial past. D-Cube City uses entertainment, cultural and compelling landscape components to drive pedestrians into activities, while further enhancing the surrounding circulation patterns. As a new landmark icon for the district, the high-rise office and hotel tower is designed by Jerde to symbolize energy growing toward the sky and the renaissance of the area as a major hub of Seoul, while referencing the site's former coal plant smoke stacks.

Nature is brought through and into the complex such as "secret" living gardens and terraces, a series of cascading indoor waterfalls, and a glass-enclosed "skylight river" sweeping through the ground level, offer an organic flow of spaces and sense of discovery. The project's sustainable design features include photovoltaic panels, grey water use for plant irrigation, geothermal heating and cooling and recycled materials for landscaping.

The cultural centre is covered with a green roof and includes a 1,277-seat performance hall and 420-seat event space with a shared lobby and outdoor garden plaza overlooking the city to the north. On multiple levels around the culture zone is a series of "Music Gardens" driven by rhythm and melody. A new public park, called Millefleur Park (Park of a 1,000 Flowers), connects the adjacent Shindorim Station to D-Cube's street-level entry, while establishing a natural connection to the Dorim River across the street.

As the landscape and pedestrian-oriented elements extend vertically, they also traverse the spaces tucked below the street level. An underground garden and event space occupy level B02 as an entertainment district floor and additional connection between the station plaza and park. As a youth-oriented area, the district houses a food garden, Korean Jang, Sweet Castle, Noodle Museum and entertainment offerings.

Inspired by the dynamic culture of Korea and the desire to re-energize the natural realm as an originating root of public culture, D-Cube City in Daesung sets a new standard in urban social activities, influencing people to evolve alongside the natural and urban environment around them.

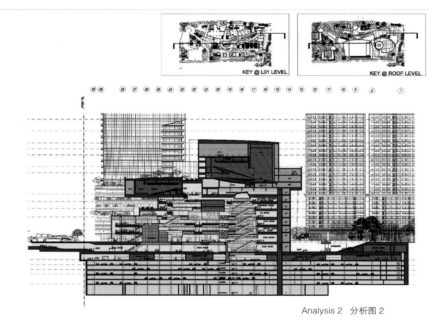

Analysis 2 分析图 2

为了在一个高密度的城市打造出一个自然和文化共存的环境，多客福城巧妙地垂直设计结合了传统韩国山水画的诸多元素。项目的设计亮点是建筑整体上像一个亚洲国家的传统灯笼。晚上，柔和的灯光透过外部覆层，吸引游客来到这里。沿着一条户外体验小路，向上通往灯笼建筑顶处的零售处；零售处以流行的当代建筑艺术展现了意大利山城特色。

首尔独具特色的多客福城，其设计中交织着自然元素，并以此来创造一个城市绿洲，重新定义该区工业城镇的历史。多客福城以娱乐、文化和引人注目的景观等元素来推动顾客流动，同时进一步优化周围的人流循环模式。作为该区一个新的里程碑式建筑，这个由捷得建筑事务所设计的集办公和酒店为一体的高层塔楼，拥有着直指天空的气概；这也是该区作为一个首尔市中心的复兴标志，同时暗指该地区曾是煤电厂烟囱的过去。

设计将自然元素完美地融入这个大型的综合体中，比如"秘密的"活动花园和露台，一系列相连的室内瀑布以及带有玻璃外层的"天际银河"倾泻而下直达地面，这些使空间变得灵动起来，给人一种渴望探索的感觉。项目的可持续设计特色包括太阳能系统、将灰水用于植物灌溉、用地热能加温或降温，还有用于景观美化的可回收材料。

文化中心覆盖着一个绿色屋顶，包括一个有1 277个座位的演艺厅、一个有420个座位的活动大厅以及能俯瞰城北的共享大堂和户外花园广场。文化中心的周围是一系列的"音乐花园"。新的"千花"花园（有1 000朵花的公园）将邻近的新道林地铁站与多客福城街道入口相连，同时与穿过大街的朵琳河自然衔接。

风景区和人行通道纵向延伸，从街道的地下空间穿梭而过。地下花园和活动大厅分布在地下二层，是专供休闲娱乐的地方，同时也成为地铁站广场和公园之间的一个纽带。作为一个青年主导区，这个分区设有美食园、蒋氏韩国料理、甜品城堡、面条博物馆和娱乐设施。

位于大成的多客福城，设计灵感来源于韩国充满活力的文化以及将自然领域重塑为公共文化根源的渴望，以此设立了一套城市社会活动的新标准来影响周边的人，让他们在自然和城市环境之间和谐发展。

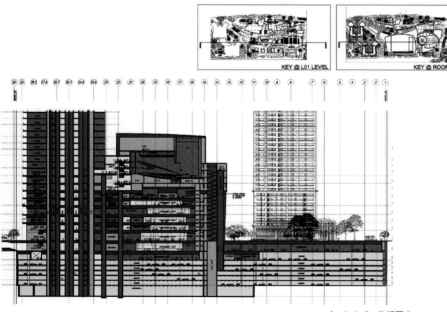

Analysis 3 分析图 3

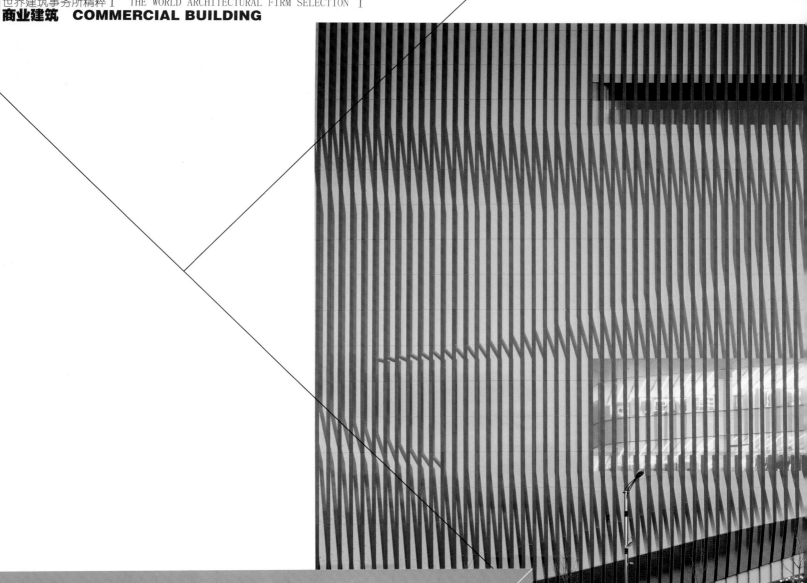

GALLERIA CENTRECITY
GALLERIA CENTRECITY 商场

Architects: UN Studio, GANSAM Architects & Partners
Project Team: Ben van Berkel, Astrid Piber with Ger Gijzen,
Marc Herschel and Marianthi Tatari, Sander Versluis, Albert Gnodde,
Jorg Lonkwitz, Tom Minderhoud, Lee Jae-young, Woo Jun-seung,
Constantin Boincean, Yuchen Lin
Location: Cheonan, Korea
Area: 110,530.73 m²

设计机构：UN 工作室，GANSAM 建筑师事务所
项目团队：Ben van Berkel, Astrid Piber with Ger Gijzen,
Marc Herschel and Marianthi Tatari, Sander Versluis,
Albert Gnodde, Jorg Lonkwitz, Tom Minderhoud,
Lee Jae-young, Woo Jun-seung, Constantin
Boincean, Yuchen Lin
项目地址：韩国天安
面积：110 530.73 平方米

The strategy for the building enclosure consists of creating an optical illusion. The facades feature two layers of customized aluminium extrusion profiles on top of a back layer of composite aluminium cladding. The vertical profiles of the top layer are straight; but those of the back layer are angled. This results in a wave-like appearance, which changes with the viewpoint of the spectator (Moiré effect).

The interior derives its character from the accumulation of rounded plateaus on long columns. The repetition of curves, enhanced by coiled strip lighting in the ceilings of the platforms, gives the interior its distinctive character. This organization has been made possible because of the complex spatial arrangement of the atrium. This central void is simple and straight in one cross-section, but oblique and jagged in the opposite cross-section. The result can be seen as a kind of spatial waterfall, a relatively narrow central void cuts through the volume from top to base, with smaller pockets of space emanating from it, like rivulets.

Four stacked programme clusters, each encompassing three storeys and containing public plateaus, are linked to the central void. This organization propels a fluent upstream flow of people through the building, from the ground floor atrium to the roof terrace. Along with the central atrium, the plateaus provide light and views both within the central space and to the exterior. As the plateaus are positioned in a rotational manner in space, they enable the central space to encompass way finding, vertical circulation, orientation and act as main attractor of the department store.

Circulation is achieved by escalators, which are positioned on both sides of the atrium; a panorama elevator by the void; and elevators placed in the east and west lobby.

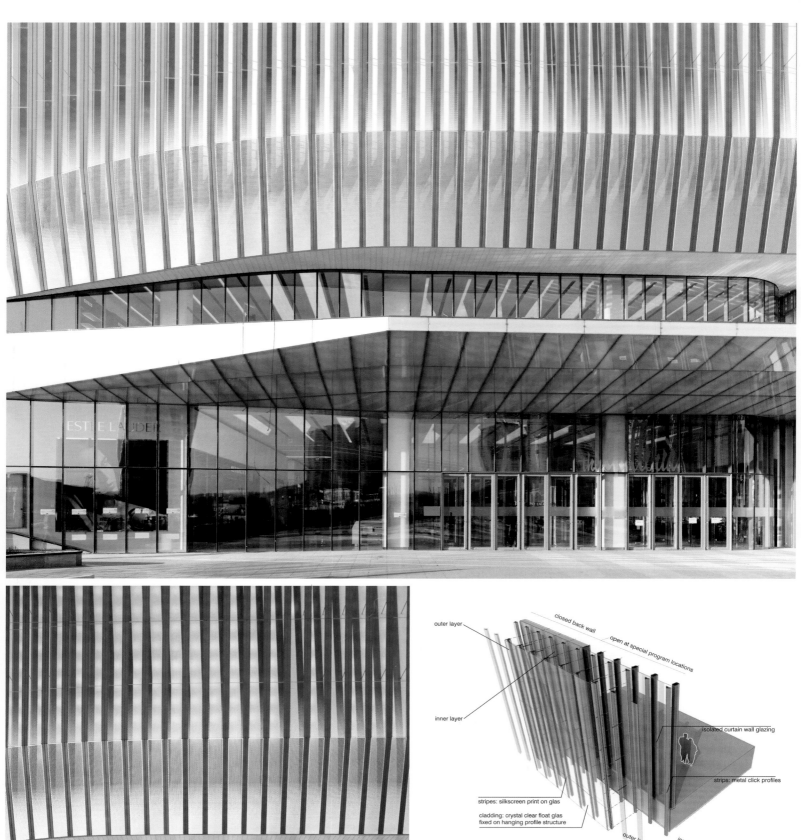

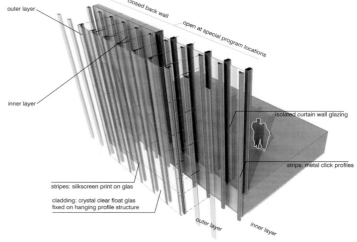

outer layer

inner layer

closed back wall

open at special program locations

isolated curtain wall glazing

strips: metal click profiles

stripes: silkscreen print on glas

cladding: crystal clear float glas
fixed on hanging profile structure

outer layer

inner layer

SPECIAL AREAS
LEVEL 09 ART CENTER

Section　剖面图

Day ➤ Night

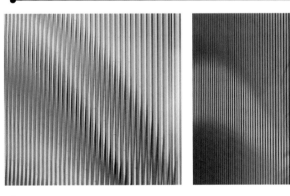

Monochrome 单色 ➤ Colored 彩色

建筑的外墙设计营造出一种视觉错觉的效果。建筑表皮是以两层特定的铝型材覆盖在复合铝合金上。顶层垂直剖面是直线形的，而下面的一层以一定角度倾斜，形成了波浪形表面。若观看的视角改变，会出现建筑的规模似乎可以改变的假象（莫尔效应）。

室内的设计亮点在于长柱上累积起来的圆形隆起。平台的天花板上，条形照明灯盘旋而下，曲线的效果得到了强化，赋予室内与众不同的特点。中庭复合式的空间布置，造就了这种组合效果。在其中一横截面上观察，中央的空间简单明了，而在与之相对的横截面上观察，空间则倾斜交错，共同构成空间瀑布的效果——由上至下，相对较窄的中央空间穿过整个建筑，零碎的小空间如同一条条的溪流以中央空间为中心散开。

四个堆积规划群，均为三层且都有公共台地，都与中央空间相连。从底层中庭到屋顶平台，这样的组合设计引导着人流自下而上地穿过建筑。台地与中庭共同提供光源，使视线可以在中央空间和建筑外部穿梭。台地旋转布置在整个空间中，它使中央空间具有路径查找、垂直循环和定向的功能，并且可以作为百货商店的主要卖点。

人流循环主要依靠位于中庭两侧的自动扶梯；中央空间有一个全景电梯；其他电梯则被设置在东大厅和西大厅。

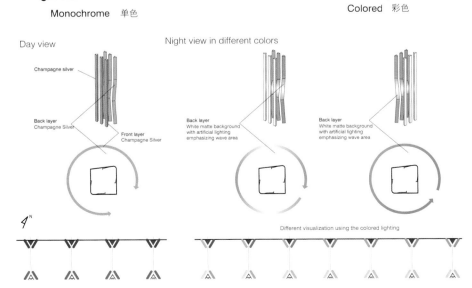

Day view

Night view in different colors

Champagne silver

Back layer
Champagne Silver

Front layer
Champagne Silver

Back layer
White matte background
with artificial lighting
emphasizing wave area

Back layer
White matte background
with artificial lighting
emphasizing wave area

Different visualization using the colored lighting

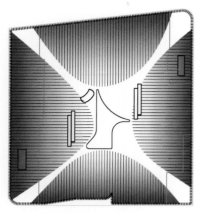

"OUTSIDE IN" FEATURE FROM SCENARIO 1

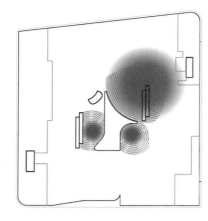

"INSIDE OUT" FEATURE OF SCENARIO 2

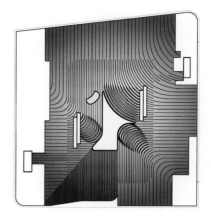

FINAL SYSTESIS

Scenario Principles　方案规划

The Galleria Cheonan Building Concept

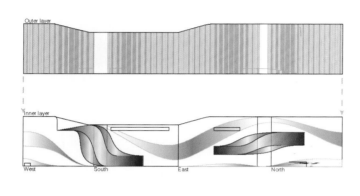

Outer layer

Inner layer

West　South　East　North

FACADE　PRINCIPLES
- ■ PLATEAU - WAVE
- ▦ GENERATED SUBWAVE
- ▢ 2 SET: OPTICAL ILLUSION

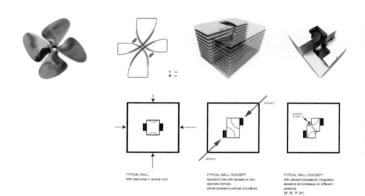

TYPICAL MALL
With balconies in central void

TYPICAL MALL CONCEPT
Applied to site with access on two opposite corners
(direct access to vertical circulation)

TYPICAL MALL CONCEPT
With stacked escalators, integrated elevators and plateaus on different positions
(5F, 6F, 7F, 9F)

North-West

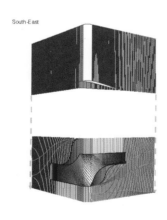

South-East

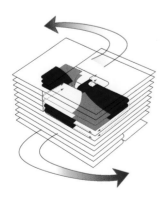

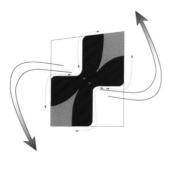

INSIDE / OUTSIDE RELATIONSHIP
- ■ PLAZA
- ▦ SPECIAL ZONE

132

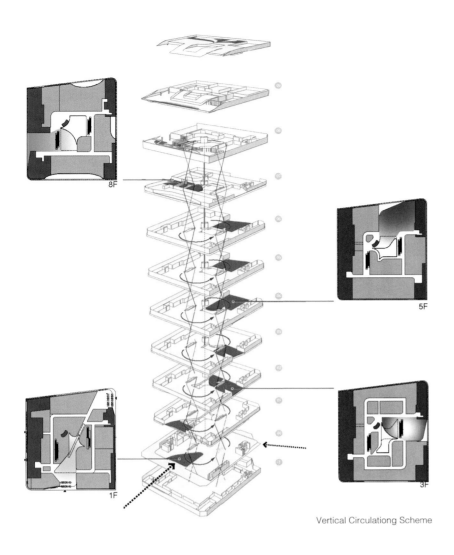

ROUTING AND PLATEAUS
The plateaus meet the routing scheme in various ponts and integrate the loop.

■ Horizontal Circulation

■ Upward Escalator

■ Downward Escalator

■ Elevator

Vertical Circulationg Scheme

Detail 细节图

DREAM DOWNTOWN HOTEL
梦想商业区酒店

Architects: Frank Fusaro, AIA, Partner, Handel Architects
Client: Hampshire Hotels & Resorts+ Vikram Chatwal Hotels
Location: New York City, America
Site Area: 2,350 m²
Building Area: 17,000 m²
Photographer: Bruce Damonte

设计机构：Frank Fusaro, AIA, Partner, Handel Architects
客户：Hampshire Hotels & Resorts+ Vikram Chatwal Hotels
项目地址：美国纽约
占地面积：2 350 平方米
建筑面积：17 000 平方米
摄影：Bruce Damonte

Dream Downtown Hotel is a 17,000 m² boutique hotel in the Chelsea neighborhood of New York City. The 12-layer building includes 316 guestrooms, two restaurants, rooftop and VIP lounges, outdoor pool and pool bar, a gym, event space, and ground floor retails.

Dream Downtown Hotel sits on a though-block site, fronting both 16th and 17th Streets, and is adjacent to the Maritime Hotel, which sits adjacent to the west. In 1964, the National Maritime Union of America commissioned New Orleans-based architect Albert Ledner designed a new headquarters for the Union, on Seventh Avenue between 12th and 13th Streets. Two years later, he designed an annex for the headquarters on the site where Dream currently sits. A few years later, Mr. Ledner designed a flanking wing for the annex, which would eventually be converted to the Maritime Hotel. In the 1970s, the Union collapsed and the buildings were sold and used for various purposes in the years that followed. In 2006, Handel Architects was engaged to convert the main annex into the Dream Downtown Hotel.

The original building offered limited possibilities for natural light. Four floors were removed from the centre of the building, which created a new pool terrace and beach, along with new windows and balconies for guestrooms. The glass bottom pool, dotted with portholes of its own, allows guests in the lobby glimpses through the water to the outside (and vice versa), and connects the spaces in an ethereal way. Light wells framed in teak between the lobby, pool and lower levels allow the guest to be transported between spaces. Two hundred handmade blown glass globes float through the lobby and congregate over The Marble Lane restaurant filling the space with a light cloud.

The otherness of Ledner's 1966 designed for the National Maritime Annex was critical to preserve. Along the 17th Street exposure, the sloped facade was clad in stainless steel tiles, which were placed in a running bond pattern like the original mosaic tiles of Ledner's Union building. New porthole windows were added, one of the same dimension as the original and one half the size, loosening the rigid grid of the previous design, while creating a new facade of controlled chaos and verve. The tiles reflect the sky, sun, and moon, and when the light hits the facade perfectly, the stainless steel disintegrates and the circular windows appear to float like bubbles. The orthogonal panels fold at the corners, continuing the slope and generating a contrasting effect to the window pattern of the north facade.

1	RESTAURANT	10	GUESTROOMS
2	LOBBY / LOUNGE	11	GUESTROOM TERRACES
3	GARDEN OPEN TO ABOVE	12	GUESTROOM PENTHOUSE SUITE
4	EVENT / EXHIBITION	13	GREENHOUSE
5	OPEN TO ABOVE	14	MECHANICAL
6	RESTAURANT KITCHEN	15	TERRACE
7	HOTEL BACK OF HOUSE	16	ROOF LOUNGE
8	LOUNGE	17	ROOF LOUNGE TERRACE
9	POOL / BEACH		

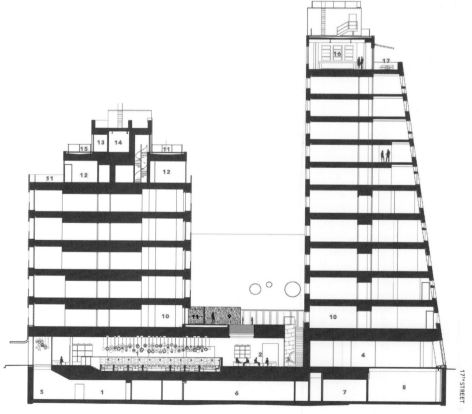

Section　剖面图

梦想商业区酒店是一家位于纽约切尔西附近的精品酒店，建筑面积 17 000 平方米。12 层的建筑包括 316 间客房、两家餐厅、屋顶花园、贵宾室、室外泳池和池边酒吧、健身房、公共空间和底层零售店。

梦想商业区酒店所在位置是一个街角，面对着第 16 和 17 大街，西面毗邻海事酒店。1964 年，美国国家海事联合会委托新奥尔良建筑师 Albert Ledner 在第 12 和 13 大街之间的第七大道为其设计一个新总部大楼。两年后，他在如今的酒店位置为总部设计了一座附楼。几年后，Ledner 又为这座附楼设计了一座翼楼，最终变身为海事酒店。20 世纪 70 年代，海事联合会解体，在随后的几年里，联合会的几座大楼先后被出售，并转做各种用途。2006 年，Handel 建筑师事务所受托，将附楼改造为梦想商业区酒店。

原先的体量不足以让自然光线随意穿透，四层楼从建筑中心迁移，使建筑的中央地带留空，沿着新客房的窗户与阳台形成了新游泳池露台和躺椅休息处。玻璃底的泳池让酒店大堂的客人能够通过水池看到外面的天空（反之亦然），使外界与建筑空间以一种空灵的方式连接。采光井镶在大堂、水池和较低楼层之间的柚木框架中，使空间富有流动性。两百个手工吹制的玻璃球体飘浮在大堂上空，与旁边的 Marble Lane 餐厅融为一体，这片神奇的"云彩"仿佛充满了整个空间。

Ledner 在 1966 年为美国国家海事联合会设计的附楼极具特色，也是需要保护的重点。沿着第 17 大街的立面成倾斜状，外表覆盖着小块的不锈钢板，采用了顺转砌合的方式，与 Ledner 设计的原联合会大楼外立面的马赛克瓷砖类似。新楼立面上的窗户如同船只的舷窗，一种与原先的大小相同，另一种为原来的一半大小，比先前的设计显得宽松，创造了一个排列有序的又充满活力的新立面。该立面白天能反射蓝天、白云和耀眼的日光，夜晚亦能反射皎洁的月光，当光线完全照射到立面上时，不锈钢表面如同破裂消失一般，而圆形窗户看起来如同漂浮的气泡。面板在建筑拐角处垂直折叠，继续保持倾斜，与北立面的窗户排布模式形成了鲜明的对比。

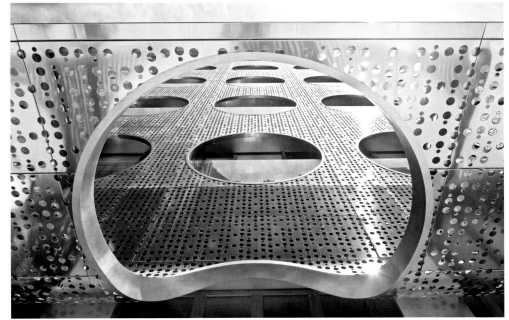

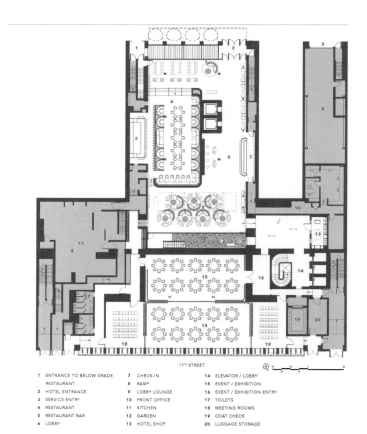

1	ENTRANCE TO BELOW GRADE	7	CHECK-IN	14	ELEVATOR / LOBBY
	RESTAURANT	8	RAMP	15	EVENT / EXHIBITION
2	HOTEL ENTRANCE	9	LOBBY LOUNGE	16	EVENT / EXHIBITION ENTRY
3	SERVICE ENTRY	10	FRONT OFFICE	17	TOILETS
4	RESTAURANT	11	KITCHEN	18	MEETING ROOMS
5	RESTAURANT BAR	12	GARDEN	19	COAT CHECK
6	LOBBY	13	HOTEL SHOP	20	LUGGAGE STORAGE

Ground Floor Plan 底层平面图

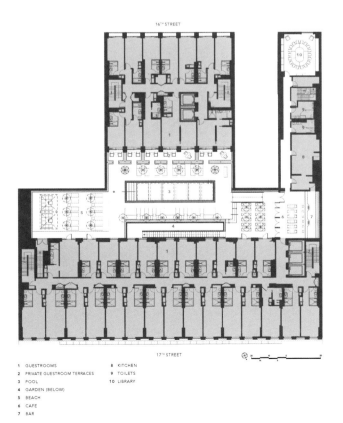

1 GUESTROOMS   8 KITCHEN
2 PRIVATE GUESTROOM TERRACES   9 TOILETS
3 POOL   10 LIBRARY
4 GARDEN (BELOW)
5 BEACH
6 CAFE
7 BAR

2nd Floor Plan 二层平面图

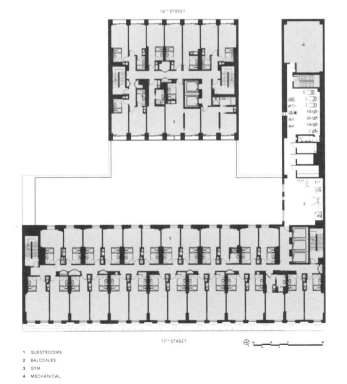

1 GUESTROOMS
2 BALCONIES
3 GYM
4 MECHANICAL

3rd Floor Plan 三层平面图

NOVA DIAGONAL TOWER
诺瓦对角线大厦

Architects: Martinez Sisternas Arquitectes i Associats (MSAA GROUP)
Developer: Vallehermoso División Promoción SAU
Construction Company: SACYR SA
Area: 17,530 m²
Height: 22 floors / 88 m

设计机构：Martinez Sisternas
Arquitectes i Associats (MSAA GROUP)
开发商：Vallehermoso División Promoción SAU
建筑公司：SACYR SA
面积：17 530 平方米
建筑高度：22 层 /88 米

This project is part of the master plan for the northern side of the extension of Avinguda Diagonal towards the sea. It forms part of an unbroken building front, in alignment with the avenue, punctuated by a series of taller structures, approximately equidistant, situated in strategic positions. The Agbar Tower by Jean Nouvel, the Hotel ME by Dominique Perrault and the Hotel Princess by Oscar Tusquets are other landmarks on this side of Avinguda Diagonal.

The tower, therefore, is not an independent object. On the contrary, it is part of the growth in height of the line of buildings, at a junction where three tall buildings coincide, overlooking an ample street section. The tower does not negate the background urban layout, made up of streets and squares defined by medium-height facades.

Although it is an independent structure, we drew up the project for the tower in conjunction with the plans for the buildings located between the streets of Fluvià and Bac de Roda, with which the tower shares expressive resources and materials.

The tower constitutes a prism, with a rectangular floor plan, aligned with the Avinguda Diagonal on its narrower side. It connects to the other buildings through a bridging structure three stories high, intended for small offices. Pedestrians can pass under this structure to access the public space inside the block, and it also covers the vehicle entrance to the car park.

The floor plan is organized around two stairwells centred at either end, connected to each other, providing access to a number of homes that vary from four to eight per floor depending on the height. The main rooms are on the street side, while the services are concentrated on the inside, next to the lifts and vertical installations. Most of the living rooms are located on the corners, and the bedrooms are aligned with the long facades, facing east or west. The double orientation of the living rooms ensures sunlight all day long.

The care taken in the definition of the volume and the approach used to design the facades, reveal a firm commitment to an architectural style of a decidedly contextual and urban nature. The facades are tense surfaces constructed of composite panelling and dark-lacquered aluminium profiles, the sole material used for frames, sun blinds and sections, in order to visually unify solid and empty space.

A white mesh, with the same pattern, is superimposed on this dark surface. It is made of lightweight, glass reinforced concrete (GRC) and forms a slightly protruding grid, emphasizing the structure and the edges of the building.

Some elements of this mesh are strategically concealed, so that they produce no effects related to order and scale which take into account views of the building from the middle and far distances. The distortions in the white mesh, in the surface and on the edges, produce an optical illusion.

The three-storey bridge connecting the tower and the adjacent line of buildings, together with the ambiguities of the incomplete mesh covering the main structure, adds interest and individuality, generating a result that is expressive and lively without ceasing to be part of a coherent building front.

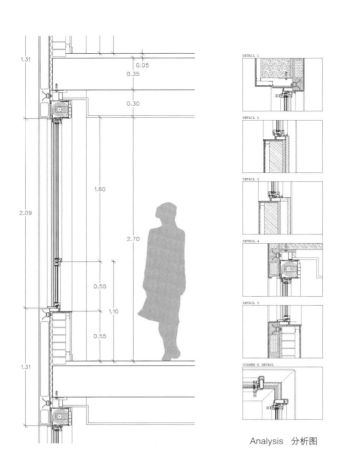

Analysis 分析图

此项目为对角线大道北面向海延伸总体规划的一部分。它是一套完整建筑群的一部分，与对角线大道成一条直线，中间穿插了一排处于重要区位、间距大致相等的高层建筑。它与让·努维尔设计的 Agbar 大厦、多米尼克·佩罗设计的 ME 酒店以及奥斯卡·图斯奎设计的公主酒店，一起作为对角线大道北侧的标志性建筑。

因此，大厦并不是一个独立的项目。相反，它是建筑群高度联排叠加的一部分，与另外三个高层建筑完美衔接，俯瞰宽敞的街面。大厦并不否定城市背景中那些典型的由中等高度幕墙围成的街道和广场。

虽然大厦是一个独立的结构，但我们的设计规划使其与位于 Fluvià 大街和 Bac de Roda 大街之间的建筑互相协调，以便共享丰富的资源和材料。

根据矩形平面图，大厦形成一个棱柱结构，与对角线大道较窄一侧成一直线。它通过一个三层高桥楼结构与其他建筑相连，用于小型办公。行人可由此通向大厦内部公共空间，同时车辆也经此进入停车场。

平面图的设计围绕着处于两端中心的两个楼梯间，两者相互连接，以此通向各个户房，依据高度，每层的户房数量由四间到八间不等。主房邻近街边，服务区都集中在内部，紧邻电梯和垂直设施。客厅大部分位于拐角处，卧室与长端幕墙排齐，面向东面或西面。而客厅的双向定位确保了一整天阳光充足。

对体积的定义以及对立面设计方法的关注，体现了对于尊重环境和城市属性这种建筑风格的坚定承诺。幕墙由复合镶板和深色涂漆铝合金型材紧密排列，单一的材料用于框架、遮阳篷和截面，以求从视觉上统一或虚或实的空间。

相同图案的白色网格叠加在深色幕墙上。它由轻质玻璃纤维混凝土构成，并形成一个稍微凸起的网格，用以突出结构和建筑边缘。

考虑到建筑中及远的视野，网格中的某些部分被明智地隐蔽起来，使它们对序列和比例毫无影响。白色网格表面和边缘的扭曲产生一种视错觉。

三层高的桥楼连接大厦与相邻建筑，连同覆盖主结构不完全的网格，增加了趣味和个性，形成一个富有表现力的生动建筑，而不再仅仅是一个衔接建筑群的单一部分。

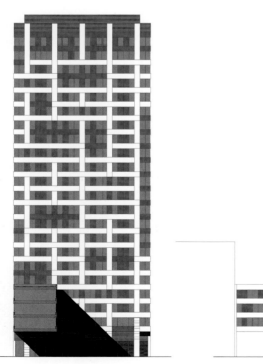

West Elevation 西立面图

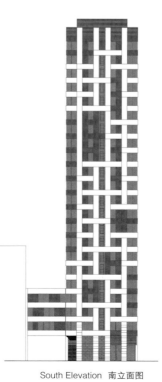

South Elevation 南立面图

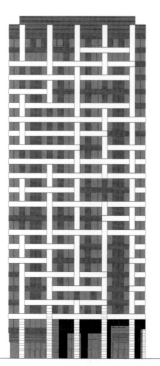

East Elevation 东立面图

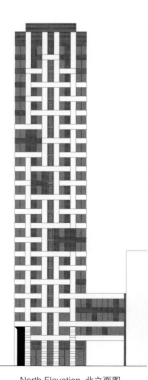

North Elevation 北立面图

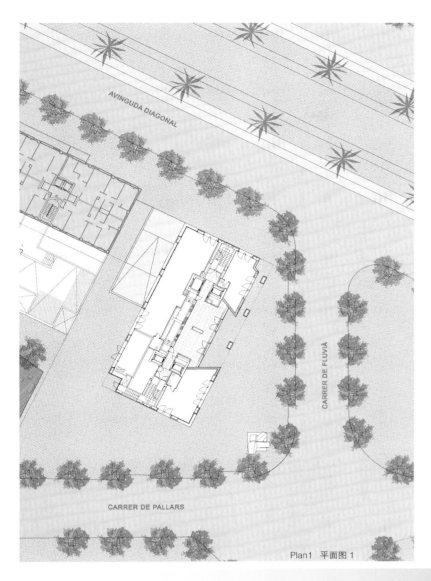

Plan1 平面图 1

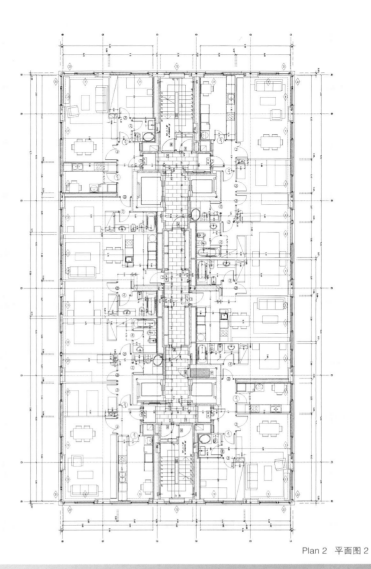

Plan 2 平面图 2

FORTE NANSHAN
长处南山

Architects: Spark Architects
Client: Forte
Project Team: Eldine Heep, Leonardo Micolta, Javen Ho
Location: Chongqing, China
Photographer: Jonathan Leijonhufvud

设计机构：Spark 事务所
客户：Forte
项目团队：Eldine Heep, Leonardo Micolta, Javen Ho
项目地址：中国重庆
摄影：Jonathan Leijonhufvud

Spark's playful clubhouse in Chongqing has opened to the public. The clubhouse is the first part of a series of leisure facilities Spark designed for a large villa community in the outskirts of Chongqing. Apart from the 5,500 m² clubhouse Spark's masterplan proposal includes a 12,000 m² hotel and a 13,000 m² outdoor retail street which will be completed early 2013.

"The uniqueness of the site was the source of inspiration for our design", says Jan Clostermann, Director of Spark. The design aims to embody the site's steep topography, beautiful greenery and scenic views in the spatial experience of the architecture. The hotel, clubhouse and retail elements become part of a scenic route from a panoramic lookout fifty meters above the community down into the valley. Along this dramatic descent program has been carefully placed to respond to the topography and community needs. Thus, in the process of meandering through the landscape, one engages in activities that foster a healthy community and social interaction.

The clubhouse forms the centre point of this leisure route. It functions as a mediating icon that brings residents, visitors, and all the elements of the master plan together.

It was formally conceived as a continuous folded loop that pivots from a common double height space. The common space links together the different programmatic elements of the building; including a gym, a swimming pool, children playground, offices, arts-and-crafts classrooms, meeting rooms, a restaurant and a cafe. The design responds to the client needs by allowing this common space to be temporarily utilized as a showroom and sales floor for the villas. Surrounding the clubhouse is a shallow reflective lake abundant with carp and lotus. Wooden platforms outline the periphery of the building and bridge across the lake connects the lobby with the residential villas.

Like the hotel and the retail street, Spark aimed to embed the clubhouse into the existing topography. There is a 5 meters level difference between the side facing the street and the side opening up to a shallow lake.

The roof folds down towards the main entry, forming outdoor continuing scenic route along the roof down towards the water. Thus the clubhouse seems to grow from the ground when approaching the main entry, and rising out of the reflecting water surface five meters lower on the opposite side of the building. The materiality aims to blend the building into its natural setting by employing textured granites and glass with simple color.

By embracing the existing topographic conditions and integrating the water edge to be one of the main features of the design, Spark has created a memorable indoor–outdoor experience for both residents and visitors alike.

 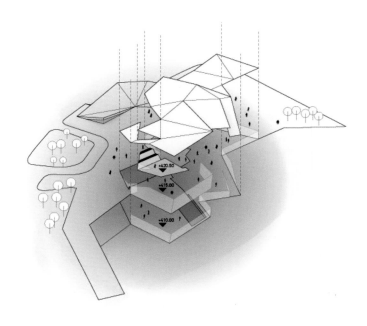

Roof Topography / Building Levels Diagram 屋面地形 / 建筑空间竖向层次

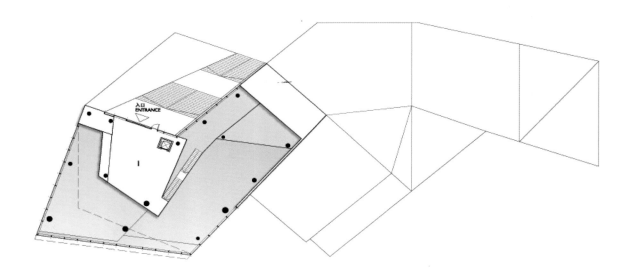

Clubhouse Level 2 Plan 会所二层平面图

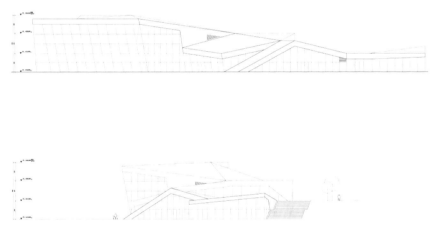

Clubhouse South East Elevations 会所南、东立面图

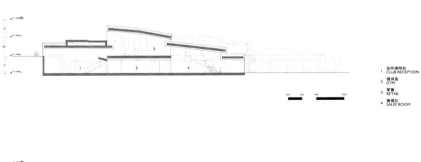

1 会所接待处 CLUB RECEPTION
2 健身室 GYM
3 零售 RETAIL
4 售楼处 SALES ROOM

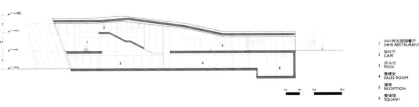

1 24小时无国别餐厅 24HR RESTAURANT
2 咖啡厅 CAFE
3 游泳池 POOL
4 售楼处 SALES ROOM
5 接待 RECEPTION
6 壁球馆 SQUASH

Clubhouse Cross Section 会所剖面图

这个由 Spark 事务所设计的俱乐部位于重庆,现已向公众开放。这个俱乐部是由 Spark 事务所为重庆郊区一个大型的别墅社区设计的一系列娱乐设施中的一部分。除了 5 500 平方米的俱乐部项目,该总体规划还包括一个 12 000 平方米的酒店和一条 13 000 平方米的户外商业街,项目整体将于 2013 年初完工。

Spark 的主管 Jan Clostermann 评论:"场地的独特性是这个设计的灵感来源。"设计的目的是为了将场地的陡峭地形、美丽的绿植和优美景色融入建筑当中。酒店、俱乐部和零售商店将成为从社区之上 50 米高的全景观景台延伸到山谷的一条风景线的一部分,沿着这条路线精心排布着回应场地及满足社区需求的项目元素。当人们徜徉在这条路上时,一种健康的社区和社会互动就此形成。

俱乐部成为休闲路线的中心点,可以将居民、游客和总体规划的所有元素集中到这里。

它像一个连续弯曲的环,以公用空间为中心旋转。公共空间将建筑的不同部分联系在一起,包括健身房、游泳池、儿童游乐场、办公室、工艺美术教室、会议室、餐厅和咖啡馆等。公共空间被安排在别墅的商业层,可以满足客户需求。会所四周是一个浅湖,湖里有很多鲤鱼和莲花。建筑四周围以木制平台,横跨湖上的小桥将大堂和住宅别墅连接起来。

就像酒店和零售街区一样,Spark 意欲将俱乐部嵌入现有地形中,建筑的临街侧和浅湖侧有 5 米高差。

屋顶朝着主入口向下折,户外景观动线沿屋顶向下延伸至水面。在接近主入口一侧,俱乐部似乎从地面生长出来,在有 5 米高差的相反一侧,俱乐部则呈现出从反光水面一跃而出的效果。设计师采用了质感粗糙的花岗岩和单色玻璃,使建筑很好地融入周围环境中。

通过充分利用现有地形和水边线的特色设计,Spark 为住户和游客创造了一个令人难忘的室内外体验空间。

THE INDEX
INDEX 大厦

Architects: Foster + Partners
Client: Union Properties
Associated Architects: Khatib & Alami
Location: Dubai, United Arab Emirates
Site Area: 20,000 m²
Height: 326 m
Photographer: Nigel Young

设计机构：Foster + Partners
客户：Union Properties
合作机构：Khatib & Alami
项目地址：阿拉伯联合酋长国迪拜
占地面积：20 000 平方米
建筑高度：326 米
摄影：Nigel Young

Located on a prominent corner site within the "Dubai International Finance Centre", the Index is a continuation of Foster + Partners' exploration of the skyscraper in the 21 century. The mixed-use components of the 80-storey, 328m tower are clearly defined, and combine 520 luxury apartments with 25 floors of office space, as well as shops, restaurants, a pool and a health club.

The distinctive form is generated by a desire to reveal the structural system and internal organisation of the tower. The upper-level block of apartments are supported by four attenuated A-frame "fins"at 27m centres, which are buttressed at each end elevation. 25 levels of column-free office accommodation are held within from these fins, separated from the residential levels by a glazed sky-lobby. The tower is orientated east to west to maximise views over the Finance Centre and to the coastline and desert beyond. This orientation also reduces solar gain, as the building's core mass absorbs heat and reduces its reliance on mechanical ventilation. A system of sunshades shelters the interiors on the exposed south elevation.

The Index sits on a generously landscaped plinth level — with sculpted pools of water and an underground car park — and is entered through a dramatic four-storey foyer. The main lift cores, which serve the office floors, are located to the east and west extremities of the tower. A small central lift core, serving 40 levels of apartments, rises to a double-height sky lobby above the office levels and provides accesses to a variety of facilities for residents, including a reception, lounge, restaurant, and fitness centre with a swimming pool. A local lift core then transports residents to their individual apartments. The tower is crowned with 12 luxurious duplex and triplex penthouse apartments with spectacular views over Dubai.

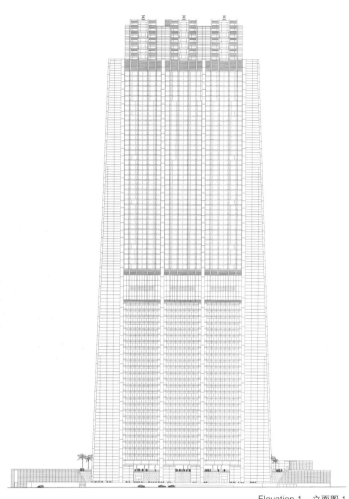

Elevation 1　立面图 1

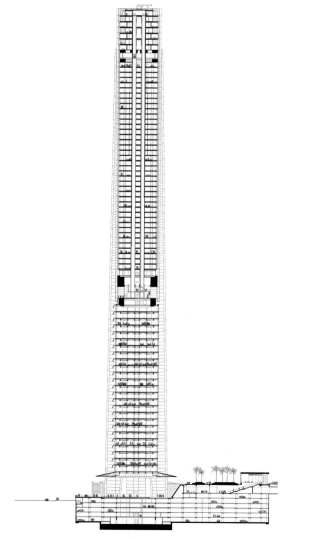

Elevation 2　立面图 2

　　Index 大厦坐落于迪拜国际金融中心内一个突出的街角，它是 Foster + Partners 建筑师事务所在 21 世纪对摩天大厦的又一探索。80 层、328 米高的多功能大厦拥有 520 间豪华公寓和 25 层的办公空间，同时包括商店、餐厅、游泳池和健身俱乐部。

　　独特的形式源于展示大厦结构系统和内部组织的愿景。高层的公寓由中心 27 米处的四个 A 形翅片支撑，这些翅片在立面端被加固。25 层的无柱办公空间通过这些翅片和玻璃的空中大厅与住宅楼层相分隔。大厦坐东朝西，将金融中心视野最大化，并将海岸线和远处沙漠的美景尽收眼底。这样的朝向可以减少太阳直射，因为大厦的中心体块能吸收热量，也可降低对机械通风的依赖。暴露的南立面有一个遮阳篷为内部遮阳。

　　Index 大厦坐落于一个大型的景观基座上，拥有一个雕塑水池和一个地下停车场。通过一个四层高的大厅可进入大厦。为写字楼服务的电梯位于大厦的东西两端。一个小型的中央电梯为 40 层的公寓服务，可以到达办公空间上方两层高的空中大厅，住户可以从此处到达接待室、大厅、餐厅和带有游泳池的健身中心。局部的电梯可以将住户送到他们各自的公寓。大厦拥有 12 套奢华的双层复式公寓和三层复式顶层公寓，可尽享迪拜的美景。

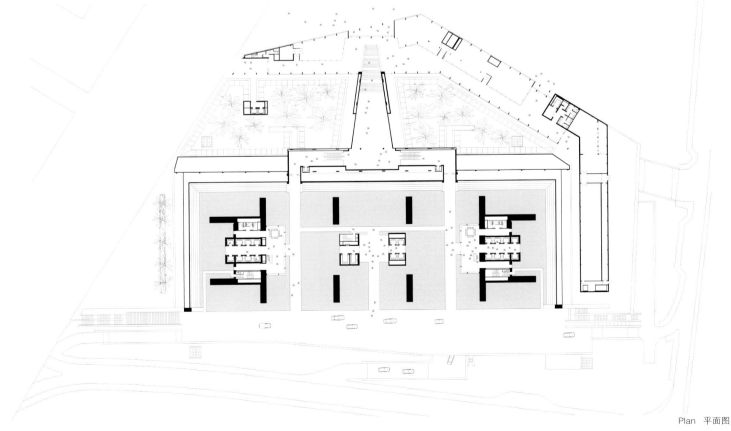

Plan　平面图

LIVERPOOL DEPARTMENT STORE
利物浦百货商店

Architects: Rojkind Arquitectos
Project Team: Michel Rojkind, Gerardo Salinas
Area: 18,000 m²
Location: Huixquilucan, State of Mexico
Photographer: Paúl Rivera

设计机构：若基柯德建筑事务所
项目团队：Michel Rojkind, Gerardo Salinas
面积：18 000 平方米
项目地址：墨西哥威斯基鲁康
摄影：Paúl Rivera

Understanding the new role shopping centres play in today's society, in which they have become a magnet for social encounters and even cultural exchanges, Rojkind Arquitectos was commissioned to design an 18,000 m² facade for the new department store as part of a new era in the company's pursuit for re-branding itself.

Liverpool department stores, with a 164-year-old history, have for the most part always been one of the main anchor stores for large shopping centres in Mexico. Its strategic location plays an important role in the immediate urban context. Located in the northern "car dependent" suburb of Interlomas on the outskirts of Mexico City, this relatively new suburb is characterized by a lack of open public space and a myriad of roads on which pedestrians are not welcomed.

The new facade responds to a fast pace of the everyday life in this isolated suburb, sitting in the middle of a very congested intersection of highways and overpasses, which give it a futuristic "Blade Runner-like" feel. With an existing circular footprint, the customization process of fabricating directly from 3D models drove the ideas behind the facade design intent. Speed became a very important factor in the way the project is experienced. Flexibility, fluidity and dynamism drove the design process.

The double-layered facade shelters the store and its users from its chaotic environment. Its sleek stainless steel machine-like exterior, is intended to evolve in a very fluid way as the intense sun bathes in throughout the day. It's a contradiction to the grit and chaos of its surroundings; a juxtaposition that becomes a new reference of this part of the city.

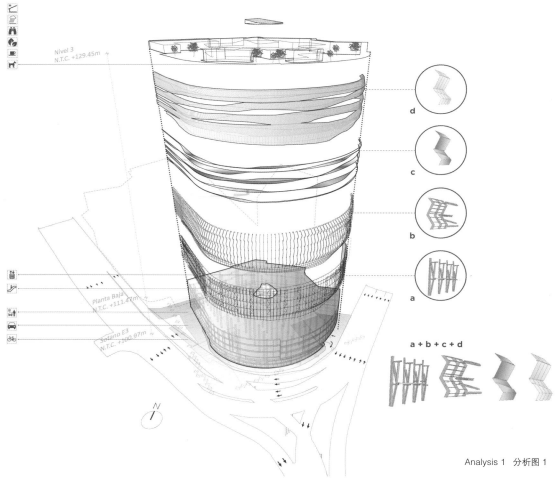

At night the hollow cavity between the layers of the facade is engulfed in light that subtly escapes through the fine reliefs formed at the folds in the skin. The facade transforms at night from its solid monochromatic appearance during the day to dynamic form accented by light.

As part of this new endeavor by the client, multiple design firms were selected to participate in the various parts of the project: the interiors were done by FRCH, the rooftop garden by Thomas Balsley and the gourmet space by JHP.

In the initial workshops sessions, it became clear that the main central interior space needed to reflect the dynamic nature of the exterior, so the client retained Rojkind Arquitectos to design this space as well.

Analysis 1 分析图 1

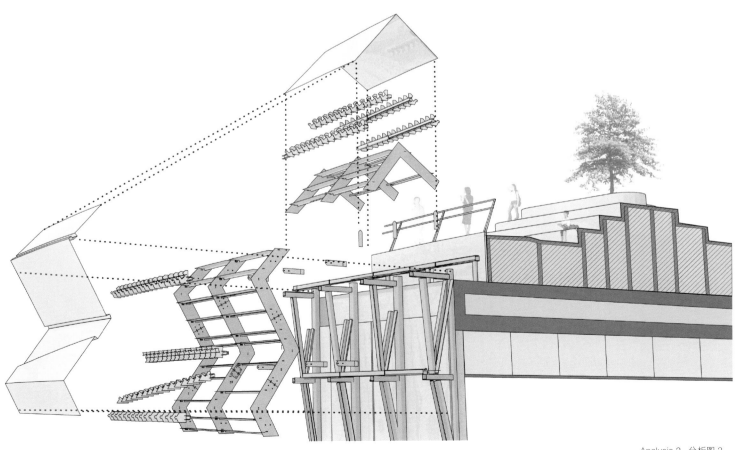

Analysis 2 分析图 2

购物中心如今已成为社会交往和文化交流的媒介，在理解新的购物中心在当今社会中所扮演的角色后，若基柯德建筑事务所被委托设计一个新百货商店的立面，面积为18 000平方米，并作为新时代里重塑公司追求的一部分。

拥有164年悠久历史的利物浦百货商店，很大程度上一直是墨西哥大型购物中心的主要核心店铺，其战略位置在快节奏的城市环境下，扮演着重要的角色。其位于墨西哥城市郊的英特罗玛斯郊区北部的"靠汽车"郊区，这个较新的郊区的特点是缺乏公共空间，还缺少大量的道路，给行人造成极大不便。

购物中心坐落于一个非常拥挤的立交桥和公路的中心，新的立面响应了在这个孤立郊区的快节奏日常生活，赋予其一种"像在刀锋上奔跑"的未来感。根据其现有的圆形旧观，直接来源于3D模型的改建定制过程，推动了隐藏在立面设计之后的设计概念的成形。规划过程中，人流循环速度成了一个非常重要的因素。灵活性、流动性和动态性共同推动了设计过程的发展。

双层的立面让整个商店和顾客远离嘈杂混乱的环境。光滑的不锈钢外表使它像个机器，全天候沐浴在强烈日光下，变化着，像是流动的液体。它与混乱的周遭形成对比，成为城市这部分的一个新的参照。

夜晚，双层立面之间的虚空淹没在灯光中，灯光巧妙地穿过表面的精美浮雕图案。立面在白天披着一层坚实的单色外衣，在晚上，却变成了灯光照射下的五颜六色的外衣。

作为委托人，对这项新工程付出了一部分努力，许多设计公司都被选来参加这项工程的各个部分：内部工程由FRCH公司完成，屋顶花园由Tomas Balsley公司完成，美食空间又是由JHP公司完成。

在最初的讨论会议中，明确了主要的内部中央空间需要反映外部动态的特点，所以委托人也保留了若基柯德建筑事务来设计这个空间。

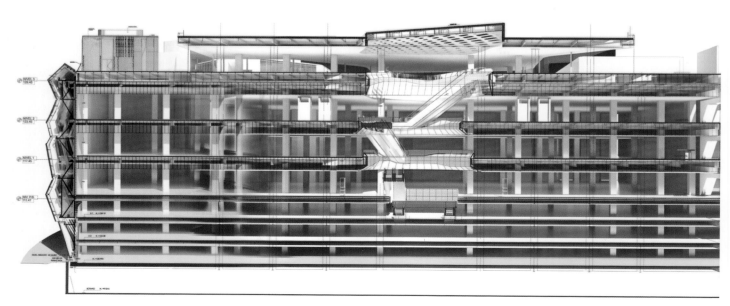

Section　剖面图

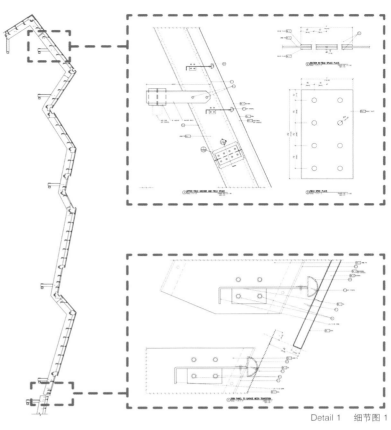

Detail 1 细节图 1

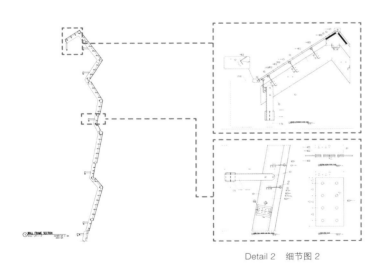

Detail 2 细节图 2

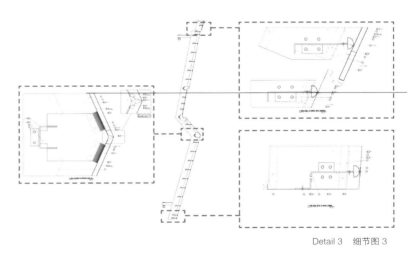

Detail 3 细节图 3

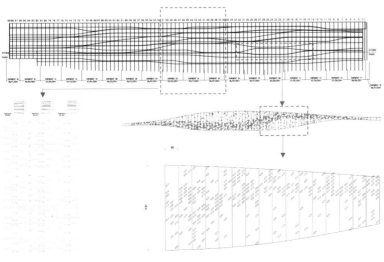

Analysis 分析图

DOLDER GRAND HOTEL
苏黎士多尔德酒店

Architects: Foster + Partners
Client: Dolder Hotel AG, Zurich
Associate Architects: Itten + Brechbuehl AG Architekten
Landscape Architects: Vetsch Nipkov Partner
Location: Zurich, Switzerland
Site Area: 400,000 m²
Photographer: Nigel Young

设计机构：Foster + Partners
客户：Dolder Hotel AG, Zurich
合作设计：Itten + Brechbuehl AG Architekten
景观设计：Vetsch Nipkov Partner
项目地址：瑞士苏黎世
占地面积：400 000 平方米
摄影：Nigel Young

Towering above Lake Zurich, the historic Dolder Grand Hotel has been reinvented to form a luxury-class city resort. The scheme integrates a substantial new extension, more than doubling the hotel accommodation and reconnecting it to the surrounding forest and resort. Remarkably, although it provides double the floor space, the new building consumes half the energy of the old or 75 percent less energy per square metre. The Dolderbahn cog railway station has been reinstated, enabling the local community to enjoy the site while experiencing something of the building itself.

The scheme restores the logic of the original hotel, designed in 1899 by Jacques Gros, and the external fabric has been restored and rendered in the original red and ochre palette. Internally, the planning has been transformed. The most significant moves have been to create a linked suite of grand public rooms, including a new ballroom, and to reinstate the grand southern entrance so that arriving guests now enjoy breathtaking views across Zurich and the Alps. Two new wings frame the historic Dolder, complementing the addition of a spa and a new ballroom.

The new wings are fully glazed,and stencil-cut aluminium screens line the facades to form balustrades and provide shading, their tree pattern resonating with the surrounding forest. While the geometry of the new elements is fluid and organic, the colour palette echoes that of the existing building to harmonise the overall composition. A highlight of the hotel is the new 4,000 m² spa. The winding stone walls that begin in the landscape continue inside to frame a canyon-like space for the pool. In some areas the walls are perforated to allow sunlight to filter in, and provide a dynamic play of light and shadow while maintaining absolute privacy. Geothermal heat pumps beneath the spa contribute to the efficient energy strategy. This is further enhanced by a high-performance envelope comprising insulated triple-glazing and natural shading.

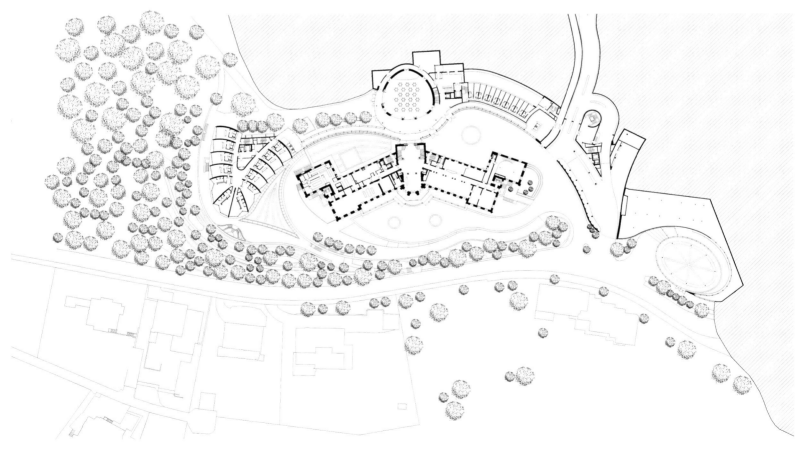

Dolder Hotel, Zurich - Ground Level

Plan 平面图

高耸在苏黎世湖上具有历史意义的多尔德酒店被改造为一个奢华的城市度假村。设计方案中不仅仅将酒店房间数量翻倍，而且对酒店进行了扩建，使其与周围的森林和度假村重新连接。引人注目的是，虽然它提供了双倍的楼层空间，但新酒店每平方米的能源消耗仅为原建筑的一半或四分之三。Dolderbahn cog 火车站已恢复使用，可以使当地居民来此享受优美的环境，体验建筑本身的独特之处。

该项目方案重新采用了原酒店（1899 年由 Jacques Gros 设计）的设计逻辑，外部肌理仍采用原来的红色和赭色。在建筑内部，设计则有所改变。最重要的改变是在大型公共空间创造连接单元，包括新的舞厅，并恢复南边的大型入口，访客在此可以享受苏黎世和阿尔卑斯山的美景。新的双翼表现了多尔德的历史，另外又新加了一个温泉和一个舞厅。

新的双翼完全铺装玻璃，模板切割的铝帘在外立面形成一排排的栏杆，并具有遮阳功能。栏杆设计为树状，与周围的森林产生共鸣。建筑新元素的流线形几何图案是有机的，和外部的颜色相呼应，使现存建筑在整体上更加协调。酒店的亮点在于4 000平方米的温泉大厅，蜿蜒的石头墙在其内部延伸形成一个峡谷般的空间。某些区域的墙壁上凿有孔洞以使阳光照射进来，在保持绝对私密的同时提供动态的光影。温泉下的地源热泵有助于有效节能。建筑的节能效果通过高品质的隔热三角玻璃外壳和遮阳系统增效后进一步增强。

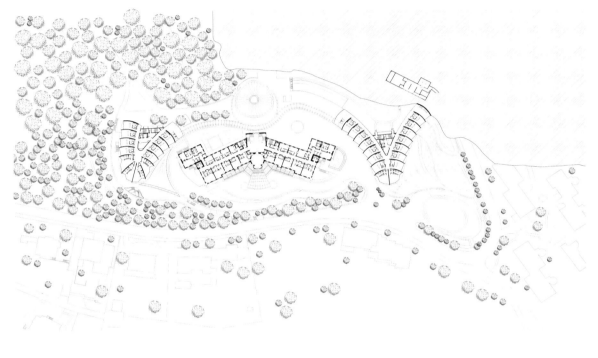

Dolder Hotel, Zurich - Level 3

Plan 1 平面图 1

Dolder Hotel, Zurich - Level 4

Plan 2 平面图 2

THE WORLD ARCHITECTURAL
FIRM SELECTION | 163

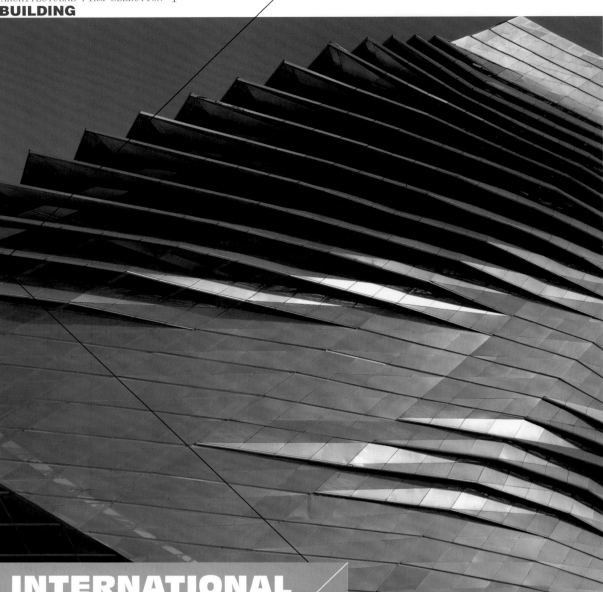

DALIAN INTERNATIONAL CONFERENCE CENTRE
大连国际会议中心

Architects: Coop Himmelb(l)au
Client: Dalian Municipal People's Government, China
Design Principal: Wolf D. Prix
Location: Dalian, China
Site Area: 40,000 m²
Building Area: 117,650 m²
Photographer: Markus Pillhofer, Coop Himmelb(l)au

设计机构：Coop Himmelb(l)au
客户：中国大连市政府
首席设计师：Wolf D. Prix
项目地址：中国大连
占地面积：40 000 平方米
建筑面积：117 650 平方米
摄影：Markus Pillhofer, Coop Himmelb(l)au

The urban design task of the Dalian International Conference Centre is to create an instantly recognizable landmark at the terminal point of the future extension of the main city axis. As its focal point the building will be anchored in the mental landscape of the population and the international community.

A public zone at ground level allows for differentiating accessibility for the different groups of users. The actual performance and conference spaces are situated at 15.3 m above the entrance hall. The grand theater, with a capacity of 1,600 seats and a stage tower, and the directly adjacent flexible conference hall of 2,500 seats, are positioned at the core of the building.

With this arrangement, the main stage can be used for the classical theater auditorium as well as for the flexible multipurpose hall. The main auditorium is additionally equipped with backstage areas like in traditional theaters and opera houses. This scheme is appropriate to broaden the range of options for the use of this space: from convention, musical, theater even up to classical opera, with very little additional investment.

The smaller conference spaces are arranged like pearls around this core, providing very short connections between the different areas, thus saving time while changing between the different units. Most conference rooms and the circulation areas have direct daylight from above.

Through this open and fluid arrangement, the theater and conference spaces on the main level establish a kind of urban structure with "squares" and "street spaces". These identifiable "addresses" facilitate user orientation within the building. Thus the informal meeting places, as well as chill-out and catering zones, and in-between the halls, gardens with view connection to outside are provided as required for modern conference utilization.

The access to the basement parking garage, truck delivery and waste disposal is located at the southwest side of the site, thus freeing the front driveway to the entrances from transit traffic. The main entrance from the seaside corresponds to the future developments, including the connection to the future cruise terminal.

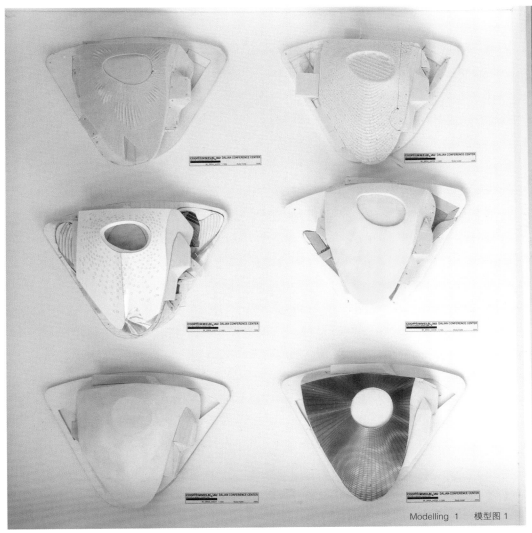

Modelling 1　模型图 1

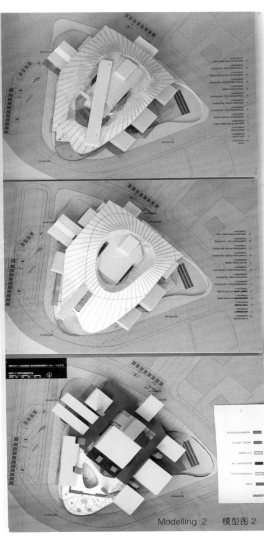

Modelling 2　模型图 2

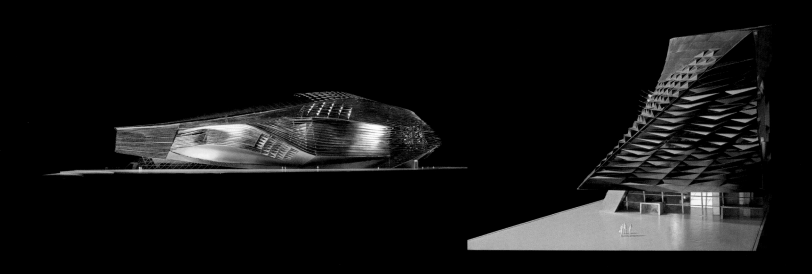

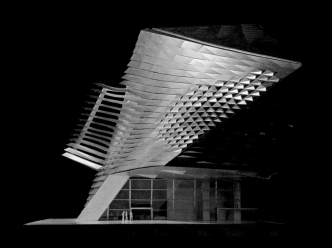

大连国际会议中心的城市规划任务是在城市主轴线的未来扩展终端上创建一个地标。作为城市的聚焦中心，此建筑将成为全体市民精神景观的象征，同时也是一个国际化的社区。

地面的公共区域为不同的用户群设置了不同的通道。执行区域和会议空间位于入口大厅15.30米以上。大剧院处于整个建筑的核心位置，拥有 1 600 个席位和一个大舞台，而与此直接相邻的灵活会议大厅拥有 2 500 个席位。

在这个设计中，主舞台可以被当做古典戏剧礼堂以及灵活的多功能大厅来使用。主礼堂额外配备了后台区域，和传统的电影院和歌剧院里的一样。此方案设计以极少的额外投资增加了空间使用方式的多样性：使其成为一个会议厅、音乐演出厅、影院甚至是古典歌剧院。

小型会议空间如同粒粒珍珠般围绕着整个大剧院，将不同的领域连接起来，为在不同空间之间转换节省了时间。大多数会议室和自由通行区域有直接的自然光从上照射下来。

通过这种开放和流动式的设计，在位于主楼层的剧院和会议空间之间建立起了一种"广场空间"与"街道空间"共存的城市结构。这些可识别的"地址"方便了用户在大楼内准确定位。本建筑应现代化会议的需要，还设置了非正式会议厅、休闲和餐饮区、中部大厅以及与外部景观相连的花园。

通往地下停车场、货车停靠站和废物处理中心的通道位于西南侧，从而使交通中转点前的车道与入口之间交通更顺畅。海上主入口与未来城市的发展相融合，其中包括通往未来邮轮码头的通道。

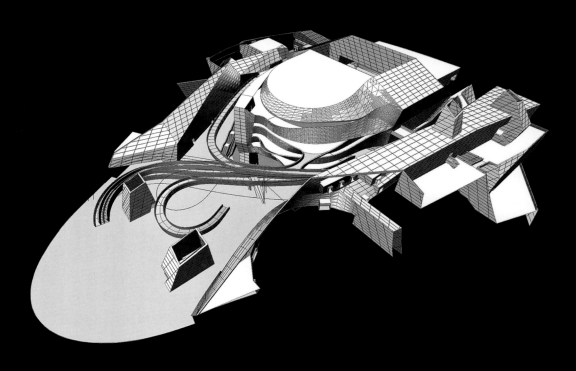

Modelling 3　模型图 3

文化建筑　CULTURE BUILDING

Being a kind of human behavior and way of life, culture is the bearer of the fruit of human civilization as well as an embodiment of the movement of civilization itself. Urban culture generally includes three parts: the material culture which represents the sensatory recognition of the city image, the technology culture which represents the spiritual identification of the city, and the norm culture which represents the regulatory distinction. It is the spirit and soul of the city. Any public cultural building can be considered as the material carrier of the spirit of the city, and the most important way to express its urban culture. However, the expression of architecture is showed by its model and space, and is melted into the architectural environment through the perception of historic culture and traditions.

When designers design public cultural buildings, they should integrate a variety of complicated factors together into a coherent whole and make a appropriate choice on the premise of comprehensive judgment. After the beforehand studies and summaries of design work, designers can begin with three aspects. Firstly, start with theme.This case usually aims at cultural displays and memorial sites for a specific topic and is always provided with relatively specific cultural performances and clear psychological anticipation on scale. Secondly, start with site environment. Many cultural buildings' cultural connotations are not provided beforehand, but depend on particular circumstances. In this case designers should proceed from the specific nature and human environments of the projects to extract useful information and manifest the true cultural connotations of buildings. Thirdly, start with contemporary values and aesthetics. With the continual appearance of new structures, new materials, new technologies, contemporary architectural forms are provided with protean manifestations. The emphasis of eco–energy saving, sustainability, merging and coexistence of public buildings and urban life, as well as sophisticated and variable building skins and rich design methods of contemporary public cultural buildings　are the most distinctive brand of the age for contemporary public cultural buildings.

Culture not only provides architecture with vitality, but also is spread widely by the expression of architecture. When building is provided with a cultural connotation, it is no longer a reinforced concrete pouring, but a cultural symbol of a city or a country.

　　文化是人类的一种行为方式和生存方式，是人类文明成果的承载，也是文明本体运动的体现。城市文化一般是由代表城市形象感观识别的物质文化、代表城市精神识别的技术文化以及代表规范识别的规范文化等组成，它是一个城市的精神与灵魂。任何一座公共文化建筑，都是城市精神的物质载体，是城市文化最重要的表达方式。而建筑的这种表达是通过造型和空间表现出来的，并通过对历史文化和传统的感悟形成认识，将之融入建筑环境之中。

　　设计师在设计公共文化建筑的时候，需要统筹考虑各种复杂因素，并在综合判断的基础上做出适宜的选择。经过对前期工作的设计研究和汇总，设计师可以从三个方面来切入：首先，以主题世界为切入点，这主要是针对某一具体的主题实践的文化展示和纪念场所而言，在这种情况下，公共建筑通常会拥有比较明确的文化性表现，对尺度也有明确的心理预期；其次，以场地环境为切入点，很多公共文化建筑所要呈现的文化内涵都不是事先给定的，这需要设计者从项目自身的具体情况出发，根据场地的自然环境和人文环境状况，提炼出有用的信息，表达出建筑所应有的文化内涵；第三，以当代价值观及审美观为切入点，伴随着新结构、新材料、新技术的不断涌现，当代建筑形式有了千变万化的表现，而强调建筑的生态节能和可持续发展，公共建筑与城市生活的共生融合以及复杂多变的建筑表皮，丰富多彩的创作设计，都为当代公共文化建筑烙下了最鲜明的时代烙印。

　　文化不仅赋予建筑以生命力，同时也通过建筑的表达而得到了加深与传播。建筑有了文化内涵，就不再是单纯的钢筋混凝土浇筑物，而是一个城市乃至一个国家的文化性标志。

CHINESE NATIONAL MUSEUM
中国国家博物馆

Architects: GMP · von Gerkan, Marg and Partners Architects
Client: Chinese National Museum
Designer: Meinhard von Gerkan and Stephan Schütz,
with Stephan Rewolle and Doris Schäffler
Landscape Design: RLA Rehwaldt Landscape Architects/Dresden, Beijing
Building Area: 191,900 m²
Location: Beijing, China
Photographer: Christian Gahl

设计机构：GMP 建筑师事务所
客户：中国国家博物馆
设计师：Meinhard von Gerkan and Stephan Schütz,
with Stephan Rewolle and Doris Schäffler
景观设计：RLA Rehwaldt Landscape
Architects/Dresden, Beijing
建筑面积：191 900 平方米
项目地址：中国北京
摄影：Christian Gahl

Design Staff: Gregor Hoheisel, Katrin Kanus,
Ralf Sieber, Du Peng, Chunsong Dong
Revised Designer: Meinhard von Gerkan and
Stephan Schütz, with Stephan Rewolle
Project Leader: Matthias Wiegelmann with
Patrick Pfleiderer

设计职员：Gregor Hoheisel, Katrin Kanus, Ralf Sieber,
Du Peng, Chunsong Dong
改建设计：Meinhard von Gerkan and Stephan Schütz,
with Stephan Rewolle
项目经理：Matthias Wiegelmann with Patrick Pfleiderer

The conversion and extension of the Chinese National Museum combines the former Chinese History Museum with the Chinese Revolutionary Museum. Completed in 1959 as one of ten important public buildings in Tiananmen Square, in direct proximity to the Forbidden City, the museum still constitutes a milestone in history of modern Chinese architecture.

Outline schemes for the conversion and extension project were selected from eleven international architectural firms, the proposal by architects von Gerkan, Marg & Partners (GMP) together with CABR of Beijing being adjudged preferred bidder. In October 2004, gmp and CABR were commissioned to build the museum, ahead of Foster and Partners, Kohn Pedersen Fox, OMA and Herzog & de Meuron, etc.

The original submission by gmp envisaged gutting the existing museum. The flying roof was planned to house the main exhibition on Chinese history, with a direct view towards the sights of the city. Following a discussion with the client and Chinese architectural experts, this scheme was revised, with the aim of integrating more of the external impact of the old building in the new building, though without abolishing the immediately obvious distinction between the old and new. This would allow the building itself to illustrate the continuity of history.

The 1959 building combines the History Museum in the south wing and the Revolutionary Museum in the north wing, separated by a central block. The task was to combine these into an integral complex of buildings by removing the central structure to make the Chinese National Museum.

The 260m-long hall acts as a central access area. It widens in the centre to embrace the existing central front entrance facing Tiananmen Square. The "forum" thus created which acts as a vestibule and multifunctional events area, with all auxiliary service functions for the public attached—cafes and tea houses, bookshops and souvenir shops, and ticket offices and toilets.

The classic tripartite division of China's historical buildings governs the design of the "forum"as well. A stone base serves as a platform for a wooden structure, with a coffered roof structure resting on a DCB layer on top of it. Despite the vastness of the "forum", a homely atmosphere was sought, particularly in the harmonious use of materials—local granite on the ground floor and walls of the base layer and wooden cladding in the gallery area.

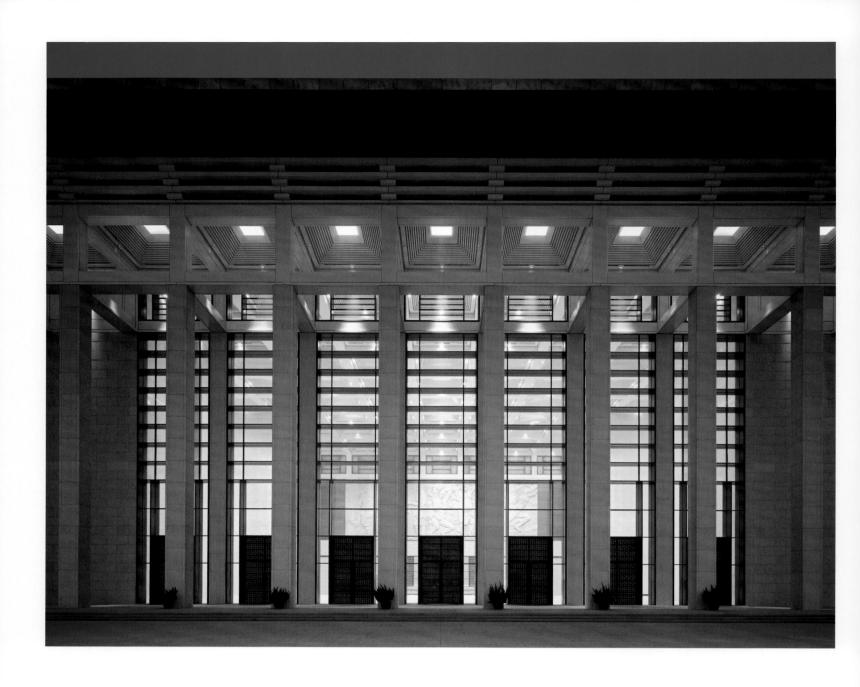

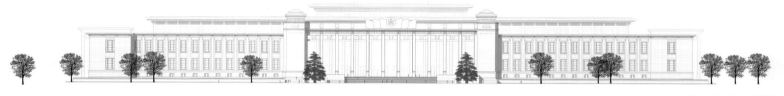

Elevation 1　立面图 1

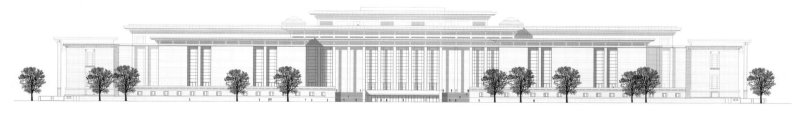

Elevation 2　立面图 2

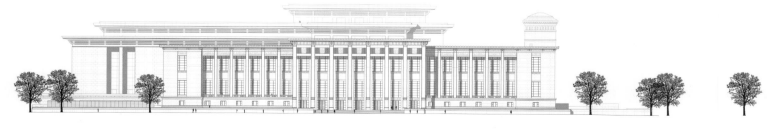

Elevation 3　立面图 3

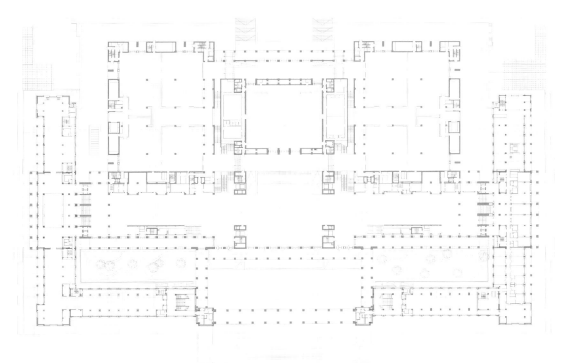

The main entrance of the museum continues to be oriented towards the west, but, for the first time, the north and south entrances are linked via the "forum". This space helps visitors to orientate, and all public areas of the almost 200,000 m² (2.066m sq. ft) building are accessed from here.

The dimensions of the "forum" also relate to the sheer size of Tiananmen Square and the size of the building itself. Around 8–10m people a year are expected to visit the National Museum. The architectural shape of the space is a contemporary interpretation of traditional elements of Chinese buildings. This is already evident in the west courtyard, accessed via broad steps reminiscent of the steps in front of the temple precincts in the Forbidden City just round the corner.

The western entrance of the existing building, the old museum facing Tiananmen Square, is notable for its series of slender pillars, linked with each other by an entablature on the pattern of temple and palace architecture, with the roof structure resting on it. The west facade of the new building is planned analogously, the "daogong" resting on its supports and carrying a prominent projecting roof. In historical Chinese architecture, the "daogong" is a slightly projecting feature of bearings and joist ends.

Plan 1 平面图 1

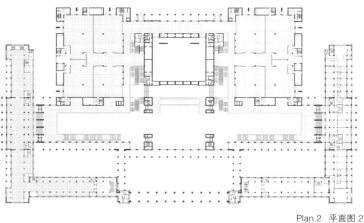

Plan 2　平面图 2

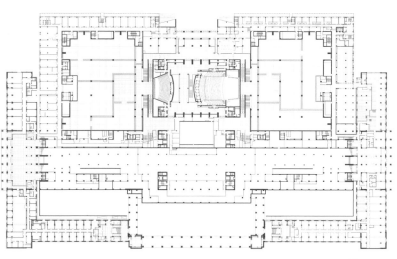

Plan 3　平面图 3

This meant the flighted roof typologies of the buildings in Tiananmen Square and the Forbidden City were continued in the new building, yet interpreted in a contemporary fashion in the detail and materials.

The entrance doors to the new buildings consist of perforated bronze plates that filter incidental daylight and thus produce a muted atmosphere in the interior, which is peculiar to traditional Chinese buildings with their ornamented window shutters. The motif of perforation was inspired by an ancient bronze panel—one of the prize items among the 1.05 m works of art that make up the National Museum collection. The ornamentation also recurs in the forming of the balustrades in the interiors of the museum.

The north wing facing Chang'an Avenue contains the exhibition relating to the modern history of China, while the south wing houses the administration and library. In the new building, the main exhibition areas are distributed over four superimposed stories, north and south of the central hall, where state receptions, banquets and similar events are held. Below the central hall is a cinema and the Academic Reporting Hall, an events room with fixed tiered seating installed, in which not only lectures but—with the planned acoustics—also classical concerts and other stage-based events can be performed. The base level and basements contain the museum's workshops and laboratories, depositories and underground garage.

The harmonious use of materials in the interiors—wood, stone and glass—is found throughout the building, creating a natural feeling of identity and familiarity. Rooms of special significance are emphasized by the use of differentiated materials. Thus the central hall opposite the main west entrance and the Academic Reporting Hall are given red wall coverings that improve the acoustics. The Jade Hall above the "forum" is notable for its backlit cast glass plates made of recycled material.

The 192,000 m² large National Museum is the largest museum in the world, its purpose being to act as a showcase for the history and art of one of the oldest cultures of mankind.

中国国家博物馆的改扩建工程结合了以前的中国历史博物馆和中国革命博物馆，它建于 1959 年，是天安门广场十大公共建筑之一，紧邻紫禁城；现在，博物馆仍然是中国近代建筑史上一个重要的里程碑。

改扩建工程的总体方案从 11 个国际性的建筑设计公司选出，由 GMP 和 CABR 提出的建设方案被认定为首选方案。2004 年 10 月，GMP 和 CABR 从 Foster and Partners, Kohn Pedersen Fox, OMA and Herzog & de Meuron 等设计公司中胜出，承建博物馆。

改建遵循了 GMP 最初的方案，将国家博物馆原有的中间体块删除，植入一个带有青铜大屋顶的新体块，用于展现中国历史，并直接享有城市视野。通过与客户和中方设计师探讨，设计师对方案进行了修改，在不破坏原有建筑风貌的基础上，使新旧建筑更加协调，这有助于建筑本身表达历史的连续性。

原建筑在南翼和历史博物馆相接，在北翼和革命博物馆相接，由中央体块分隔。设计任务是将建筑群统一成一个整体，使中央结构也成为国家博物馆的一部分。

一个 260 米长的入口大厅在中央变宽，以环抱面对天安门广场的中央前入口。这里的大厅充当前厅和活动区，具有多种功能：咖啡厅、茶馆、书店、纪念品商店、售票处和厕所。

平面布局沿用中国传统建筑的三段式布局。石质基座上是木质结构，屋顶结构则内嵌有 DCB 控制块。虽然大厅开阔，却也有家的氛围，地面和墙面的材料是花岗岩，在走廊区域则大面积运用了木材。

博物馆的主入口朝西设置，建筑师也在北部和南部开辟入口，以方便游客可以在这个近 200 000 平方米的公共场所自由参观。

大厅规模与天安门广场和建筑本身相适应，每年可接待八百万到一千万游客来访。建筑形态包含中国建筑的传统元素，这点可以从西园完全看出，园中宽阔的台阶和紫禁城里的台阶相似。

老馆建筑面对天安门广场的原建筑的西入口，因其一系列纤细的支柱而闻名，这些支柱通过檐部彼此相连，其上的图案与寺庙和宫殿建筑的相类似，上面覆以屋顶结构，建筑的西立面也与上述结构类似，上面设有倒拱，显著地突出了屋顶特色。在中国的建筑史上，倒拱是一种支座和端点托梁结构。

紫禁城和天安门的屋顶类型在这里得到延续，同时在细节和材料上又引入了当代时尚元素。

入口大门上安装了雕花青铜板，可以透过阳光，使内部氛围更加柔和，就像中国传统建筑里边的百叶窗。这种雕花灵感源于中国古代的青铜板——一块国家博物馆里高 1.05 米的珍贵的艺术收藏品。同时这种风格的装饰也出现在博物馆的栏杆细部中。

博物馆的北翼邻近长安街，主要展览中国近代史展品，南翼则是图书和管理功能区域。新增建设四个交错展览区以及南北各一个中央大厅，可以用来进行接待、举行宴会等。在中央大厅下面有一个电影院、学术报告厅、有固定座位的活动室，可以用来授课、举办音乐会和其他多种活动。地下室还有博物馆工作间、实验室、寄存处和车库等。

在内部，建筑师将木材、石材和玻璃和谐地运用在一起，非常自然、大气，并且亲切。贵宾室采用了对比性强的材料。西大厅对面的中央大厅以及学术报告厅均设计成红色的墙面，有助于改善空间音效。玉器馆因其再生材料制作的背投玻璃板而引人注目。

这个 192 000 平方米的博物馆是世界上最大的博物馆，是拥有人类最古老文明之一的历史和艺术的展示窗口。

SHANGHAI ORIENTAL SPORTS CENTRE
上海东方体育中心

Architects: GMP · von Gerkan, Marg and Partners Architects
Client: Shanghai Sports Bureau
Project Leader: Chen Ying
Structural Engineer: schlaich bergermann und partner – Sven Plieninger
Landscape Design: WES & Partner
Chinese Partner Practice: SIADR, Tongji Design Institute
Location: Shanghai, China
Gross Floor Area: 163, 800 m²
Photographer: Marcus Bredt

设计机构：GMP 建筑师事务所
客户：上海市体育局
项目经理：Chen Ying
结构工程师：schlaich bergermann und
partner —Sven Plieninger
景观设计：WES & Partner
中国合作伙伴：上海建筑设计研究院有限
公司，同济大学设计研究所
项目地址：中国上海
建筑面积：163 800 平方米
摄影：Marcus Bredt

Designer: Meinhard von Gerkan and
Nikolaus Goetze with Magdalene Weiss
Project Team: Jan Blasko, Lü Cha, Lü Miao, Jörn
Ortmann, Sun Gaoyang, Yan Lüji, Jin Zhan, Fang
Hua, Martin Friedrich, Fu Chen, Ilse Gull, Kong Rui, Lin
Yi, Katrin Löser, Ren Yunping, Alexander Schober, Nina
Svensson, Tian Jinghai, Zhang Yan, Zhou Yunkai, Zhu Honghao

设计师：Meinhard von Gerkan and Nikolaus Goetze with
Magdalene Weiss
项目团队：Jan Blasko, Lü Cha, Lü Miao, Jörn Ortmann,
Sun Gaoyang, Yan Lüji, Jin Zhan, Fang Hua, Martin Friedrich, Fu Chen,
Ilse Gull, Kong Rui, Lin Yi, Katrin Löser, Ren Yunping, Alexander Schober,
Nina Svensson, Tian Jinghai, Zhang Yan, Zhou Yunkai, Zhu Honghao

The Shanghai Oriental Sports Centre (SOSC) was built on the occasion of the 14th FINA World Swimming Championships in Shanghai. The sports complex was designed and built by architects von Gerkan, Marg and Partners (gmp), who won the competitive bidding in 2008, and constructed it in two and half years. It consists of a hall stadium for several sports and cultural events, a natatorium (swimming hall), an outdoor swimming pool and a media centre. In keeping with a sustainable urban development policy, the SOSC was built on former industrial brownfield land along the Huangpu River. The individual venues are designed so that after the Swimming Championships, they can be used for a variety of other purposes.
Water is the overarching theme of both the park and the architecture of the stadiums and the media centre. It is the connecting element between the buildings, which stand on raised platforms in specially constructed lakes. Thus the round stadiums have a curved lakeside shore round them, while the rectangular natatorium has a straight lakeside shore. Design affinities and a shared formal idiom and use of materials give the three stadiums structural unity. The steel structures of broad arches with large-format triangular elements made of coated aluminium sheet form double-sided curved surfaces along the frame of the sub-structures, thus evoking sails in the wind.

Hall stadium

During the World Swimming Championships, pool events and synchronized swimming championships took place in the hall stadium, which later can be used for boxing matches, basketball, badminton or ice-hockey matches and concerts. The hall has a crowd-capacity of 14,000, which can be increased to 18,000 by the use of mobile seating. The main structure of the closed building with a round ground plan consists of reinforced concrete, while the roof is a steel structure with a 170 m span with aluminum cladding. The parallel steel girders create 35m-high arcades and include the glass facades of the encircling open foyer.

Natatorium

The natatorium contains four pools arrayed in a row: two standard-sized, one for diving and a leisure pool. It has over 3,500 fixed seats, which will be expanded to 5,000 for the world championships, to meet FINA's requirements.

The swimming hall is a closed building with a rectangular ground plan, a main structure of reinforced concrete and a roof structure of sectional steel girders. The roof structure with triangular glass surfaces is around 210 m long, 120 m wide and 22 m high. Direct, intrusive sunlight is forestalled by means of narrow top lights along the beams, without preventing natural day lighting.

Outdoor pool

This swimming complex is located in the open on an artificial island and offers 2,000 fixed stadium seats. For the World Swimming Championships and other outstanding events, capacity is increased to 5,000 seats. The competition-size diving pool and diving towers are complemented by a competition pool.

As in the other stadiums, the roof structure with its external diameter of 130 m reflects the round ground-plan of the shell of the building. The inner diameter is around 90 m. The roof trusses are carried by the building structure. A lightweight membrane between the modules provides protection against sun and rain.

Media center

The 80 m high high-rise building is on the northern side of the sports complex. Its 15 floors include a fitness center, conference rooms and medical care centre, plus VIP and office areas. Because of the even 8.4 m grid, the building can be used flexibly. With its external shell of white, perforated aluminum panels, the building interprets the undulating shape of the adjacent lake.

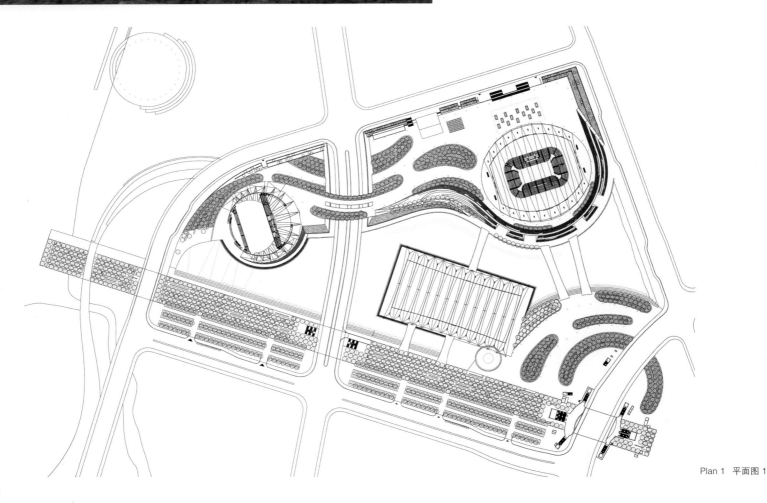

Plan 1 平面图 1

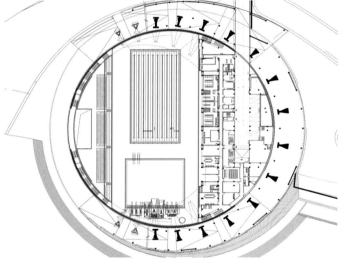

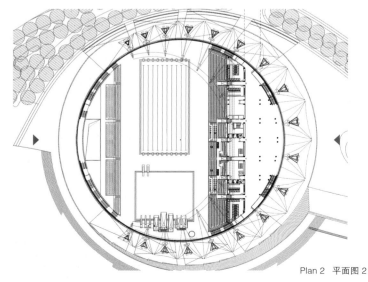

上海东方体育中心（SOSC）是为在上海举办的第十四届世界游泳锦标赛而建的。体育场馆由 GMP 设计建造，该机构在 2008 年此项目中竞标成功，场馆的建造花费两年半的时间。它包含一个能举办各种体育和文化活动的大厅式体育场馆、一个游泳场馆、一个室外泳池和一个媒体中心。为了贯彻城市发展的可持续政策，上海东方体育中心建造在黄浦江沿岸的一个废旧工业用地上。单独的场馆设计是为保证在锦标赛结束后场馆能有多方面的用途。

水是这个项目所有结构的主导性元素，也是不同结构之间的连接性元素，它出现在人工湖高抬的平面上。圆形的场馆有一个弧形的湖水边线环绕，而矩形的游泳池则是由一条直线形的湖水边线环绕。设计上的相似性、共同的主题和相似材料的使用使三个场馆有着结构上的统一性。巨大的带有大型三角形元素的铝制拱形钢铁构架沿着附属结构形成双面的弧形表皮，使人们想到了随风飘曳的风帆。

体育馆

在世界游泳锦标赛期间，游泳赛事和花样游泳锦标赛在体育馆举行，以后可用于举行拳击赛、篮球赛、羽毛球赛、冰球赛和音乐会等。大厅可容纳 14 000 人，通过使用可移动座位可增加至 18 000 人。这个圆形封闭式建筑的主要结构是钢筋混凝土，而屋顶是钢结构，并有一个 170 米宽的铝覆层。开放式前厅是平行钢桁架结构的 35 米高的拱廊，四面是玻璃幕墙。

游泳馆

游泳馆包含四个排成一行的游泳池：两个标准泳池、一个潜水池和一个休闲池。这里有 3 500 个固定座位，在世锦赛期间，座位可按需增加到 5 000 个，能达到国际泳联的要求。

游泳馆是一个矩形封闭建筑，主结构是钢筋混凝土，屋顶结构是钢桁梁。屋顶结构采用三角玻璃表层，其长约 210 米，宽约 120 米，高约 22 米。沿梁设置的灯光系统不会阻碍自然采光。

室外游泳池

游泳综合体坐落在一个人工岛上，可以提供 2 000 个固定座位。在世锦赛和其他大型体育赛事期间，座位可以增加到 5 000 个。此处还有竞赛标准的潜水池和跳台。

与体育馆的屋顶结构一样，本建筑屋顶的直径为 130 米，内径为 90 米。屋顶桁架由建筑结构支撑。模块之间的薄膜可以阻挡阳光和雨水。

媒体中心

在体育中心北侧有一座 80 米高的 15 层建筑，包括一个健身中心、会议室、医疗保健中心、贵宾区和办公区。由于采用了 8.4 米的网格结构，使其拥有很强的灵活性。建筑外层采用白色的多孔铝板，映射了周围湖水的起伏。

Plan 2　平面图 2

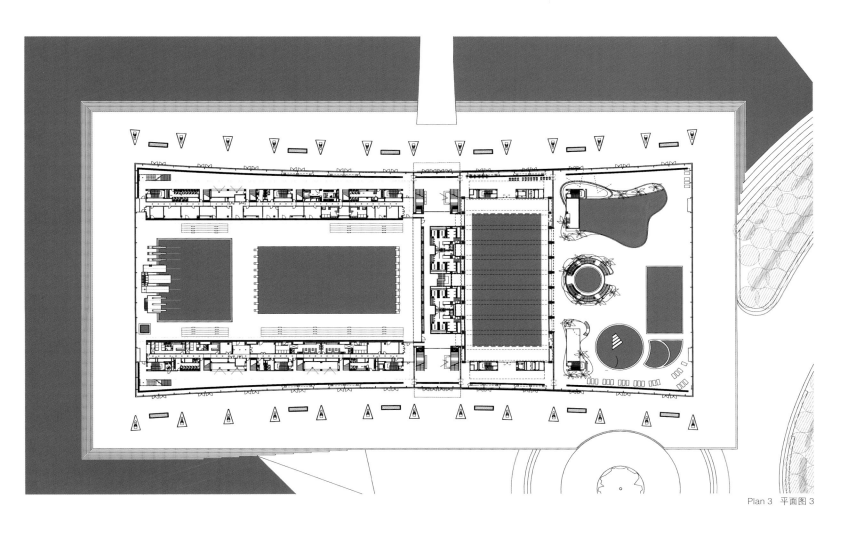

Plan 3　平面图 3

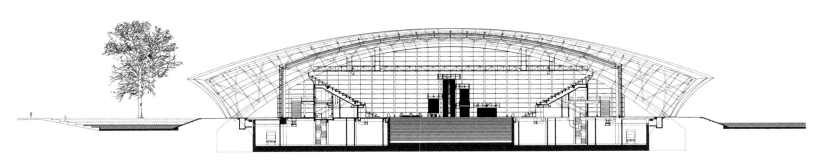

Section　剖面图

WUXI GRAND THEATRE
无锡大剧院

Architecture and Interior Design: PES-Architects
Client: Office for the Important Urban Projects in Wuxi
Location: Taihu New City, Wuxi, P.R. China
Building Area: 78,000 m²

建筑设计与室内设计：PES 建筑设计事务所
客户：无锡市城市重点工程建设办公室
项目地址：中国无锡太湖新城
建筑面积：78 000 平方米

In 2008, PES-Architects won the first prize in the invited international architectural competition for Wuxi Grand Theatre. The other competitors were established and well-known practices from Germany, France, Japan and Denmark.

The main idea of Wuxi Grand Theatre is based on its location. The man-made peninsula on the northern shore area of Taihu Lake and the highway bridge nearby make this location comparable to that of Sydney Opera House. Due to this location the building is an impressive landmark, rising up to a total height of 50 m like a big sculpture from the terraced base. Its eight gigantic roof wings stretch far over the facades, giving the building a character of a butterfly, while protecting the building from the heat of the sun.

The architectural concept is unique: inside the steel wings are thousands of LED lights, which make it possible to change the colour of the wings according to the character of the performances. This is possible, because the underside of the wings is covered by perforated aluminium panels. Another special feature is the "forest" of 50 light columns, each 9 m high, which start from the main entrance square, support the roof of the central lobby and continue outside of the lakeside entrance into the lake.

There is a strong Chinese feature that runs throughout the whole building: the large scale use of bamboo which is both a traditional and a modern Chinese material. Recently new methods for the production and use of bamboo have made it possible to cover the Main Opera Auditorium with over fifteen thousand solid bamboo blocks, all individually shaped according to acoustic needs and architectural image. There is also a material with a Finnish character: almost twenty thousand specially designed glass bricks cover the curved wall of the opera auditorium in the lakeside lobby. Finnish nature, lakes and ice, were the architectural inspiration.

Earlier references of PES-Architects international projects include St Mary's Concert Hall in Germany and Helsinki-Vantaa Airport in Finland. In China, a mix-use 192m high-rise building is currently under construction in the city of Chengdu, the result of a first prize in an international architectural competition in 2009.

To emphasize the experience of being 8 years in China design and construction market, PES-Architects in Autumn 2011 opened the Chinese design company PES-Architects Consulting (Shanghai) Co. Ltd.

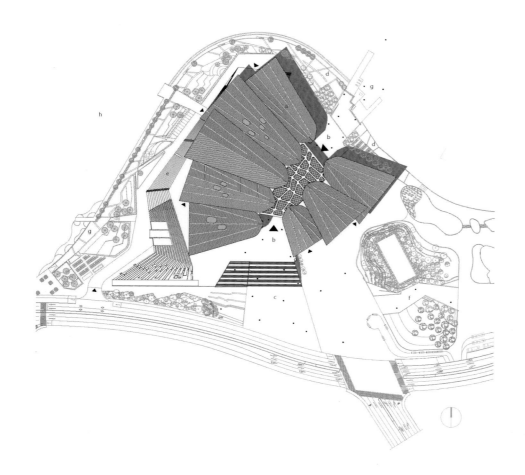

a. 8 wings, zinc roofing
b. main entrances
c. entrance plaza and main entrance steps
d. boat piers
e. green steps
f. metro station and landscape hill
g. lake-side promenade
h. wuxi lake

1:2000

Site Plan 总平面图

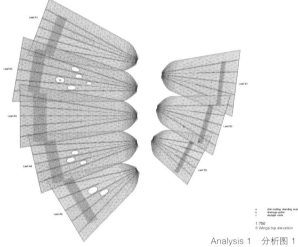

Analysis 1　分析图 1

8 Wings steel structures illustration

Analysis 2　分析图 2

2008 年，PES 建筑设计事务所赢得了无锡大剧院建筑设计国际竞赛的一等奖。其他参赛者是来自德国、法国、日本与丹麦的知名机构。

无锡大剧院的主要设计理念是基于它的区域位置。剧院选址在太湖南岸地区的人造半岛上，附近有一座高速公路大桥，这使得该基地位置可与悉尼大剧院相媲美。该位置决定大剧院成为重要的地标性建筑，建筑如同巨大的雕像一般从基地高度升起到 50 米高处。建筑的 8 个巨大的屋顶如翅膀般远远伸出墙外，让剧院看起来像一只蝴蝶，大屋顶同时也有利于阻挡阳光。

建筑的设计理念是独一无二的：在钢翅膀内是上千只 LED 灯，可以根据演出剧目的特点变换翅膀的颜色。翅膀底面覆盖的穿孔铝板让此概念得以实现。另一个特别之处是由 50 根高 9 米的光之柱形成的"森林"，光柱从主入口的广场开始，支撑起中央大厅的屋顶，并通过沿湖的入口延伸至湖中。

中国特色贯穿于整个建筑：大量使用既是中国传统材料也是现代材料的竹材。一些新的竹材制造和使用方法，使得主要剧院观众厅被15 000 块根据音响需要与建筑形象决定的各自不同形状的实心竹材块包裹成为可能。

剧院还采用了一项具有芬兰特色的材料，即剧院观众大厅沿湖一侧的弧形墙上将近两万块经过特别设计的玻璃砖。芬兰的风景、湖泊与冰是建筑设计的灵感。

PES 建筑建筑设计事务所之前的作品包括了位于德国的圣玛丽音乐厅和位于芬兰的赫尔辛基万塔国际机场在内的众多国际项目。在中国的成都市，一座 192 米高的多功能高层建筑正在建造中，PES 建筑设计事务所在 2009 年该项目的国际建筑设计竞赛中获得了一等奖。

为了加强 PES 在中国已有的 8 年的设计经验与建设市场，PES 建筑事务所于 2011 年秋季创立了在中国的设计公司，即萨米宁希诺宁建筑设计咨询（上海）有限公司。

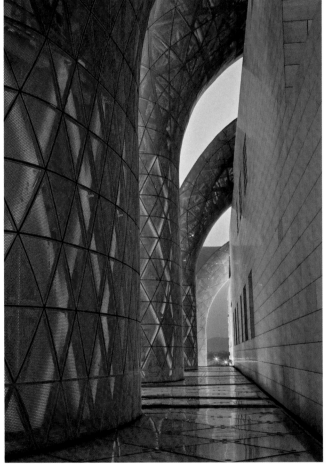

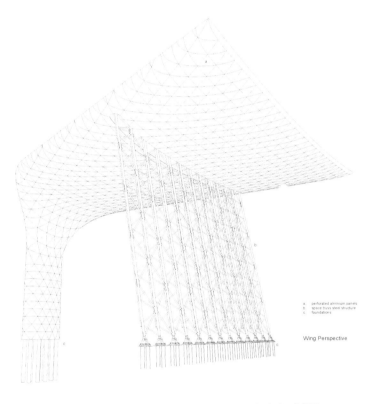

a. perforated aluminum panels
b. space truss steel structure
c. foundations

Wing Perspective

Analysis 分析图

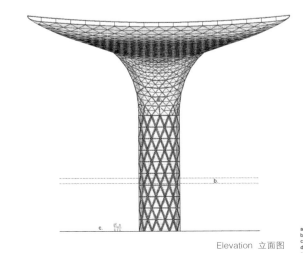

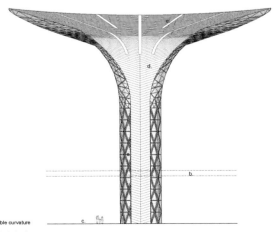

a. perforated aluminium panels, double curvature
b. intersection with foyer glass roof
c. main entrance foyer stone floor (at +6m level)
d. zink roofing, standing seam
e. drainage gutter

Elevation 立面图

Section 剖面图

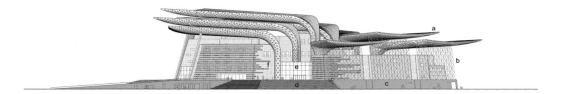

a. wing roofs
b. space truss steel wall
c. building plinth
d. main entrance steps
e. main entrance
f. stone building facede

1:750
South Elevation
南立面图

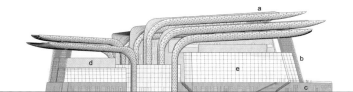

a. wing roofs
b. space truss steel wall
c. building plinth
d. stone building facade
e. glass facade

1:1500
North Elevation
北立面图

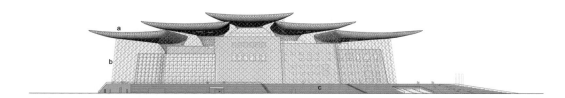

a. wing roofs
b. space truss steel wall
c. building plinth

1:1500
West Elevation
西立面图

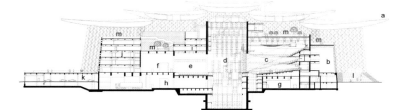

a. wing roofs
b. public foyer
c. main auditorium
d. stage tower
e. backstage
f. assembly
g. orchestra lounge
h. assembly and loading
i. rehearsal rooms
j. office / dressing rooms
k. parking hall
l. outdoor theater
m. public roof terraces

1:1500
Section
剖面图

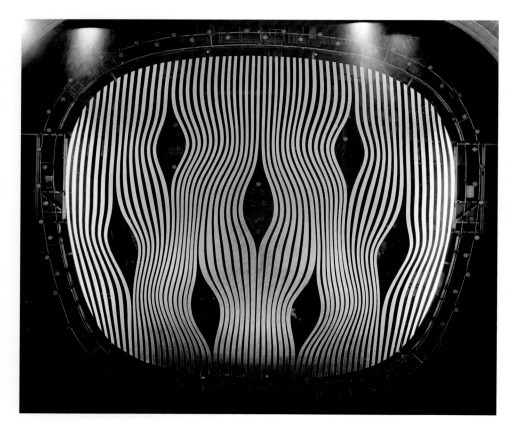

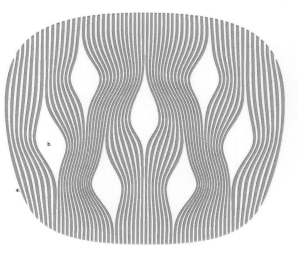

a. metal stripes, coated with leafgold
b. openings for stage lighting

1:100
Opera Mask acoustic ceiling reflector

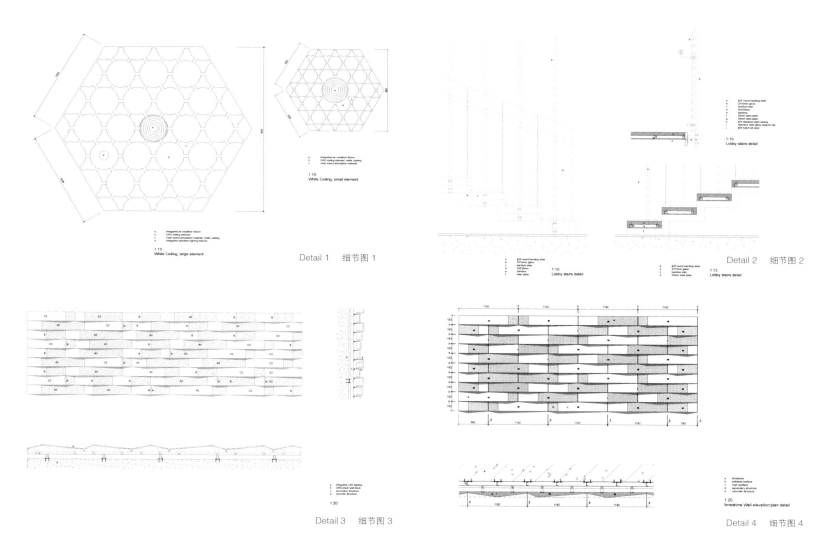

Detail 1　细节图 1

Detail 2　细节图 2

Detail 3　细节图 3

Detail 4　细节图 4

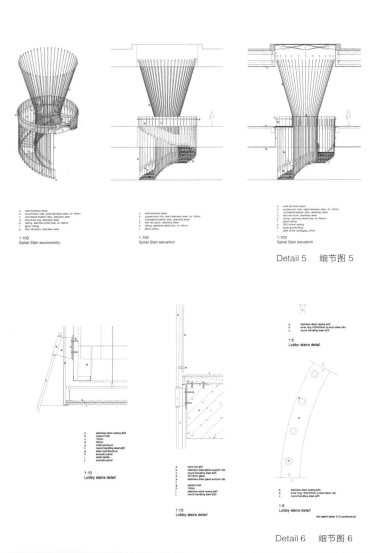

a. solid bamboo steps
b. suspension rods, solid stainless steel, d= 16mm
c. concealed fixation clips, stainless steel
d. structural ring, stainless steel
e. stair structure, stainless steel
f. railing, stainless steel tube, d= 48mm
g. glass railing

1:100
Spiral Stair axonometry

a. solid bamboo steps
b. suspension rods, solid stainless steel, d= 16mm
c. concealed fixation clips, stainless steel
d. stair structure, stainless steel
e. railing, stainless steel tube, d= 48mm
f. glass railing

1:100
Spiral Stair elevation

a. solid bamboo steps
b. suspension rods, solid stainless steel, d= 16mm
c. concealed fixation clips, stainless steel
d. stair structure, stainless steel
e. racing, stainless steel tube, d= 48mm
f. glass railing
g. GRG white ceiling
h. black granite floor
i. steel sheet cladding, white

1:100
Spiral Stair elevation

Detail 5   细节图 5

a. stainless steel casing ∅45
b. inner ring 100X40mm to bind steel rids
c. round handling steel ∅25

1:5
Lobby stairs detail

a. stainless steel casing ∅45
b. expand bolt
c. 10mm
d. 20mm
e. white structurum
f. round handling steel ∅25
g. steel roof structure
h. acoustic panel
i. white textile
j. acoustic panel

1:10
Lobby stairs detail

a. hand rail ∅50
b. stainless steel glass support clip
c. round handling steel ∅25
d. 20X10mm glass
e. stainless steel glass support clip

1:10
Lobby stairs detail

a. stainless steel casing ∅45
b. inner ring 100X40mm to bind steel rids
c. round handling steel ∅25

1:5
Lobby stairs detail

Detail 6   细节图 6

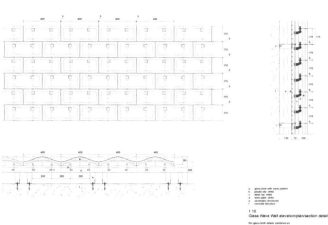

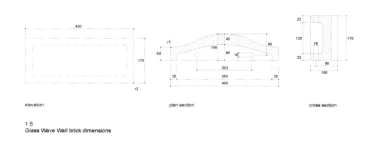

elevation

1:5
Glass Wave Wall brick dimensions

plan section

cross section

a. glass brick with wave pattern
b. plastic clip white
c. steel cap white
d. steel plate white
e. secondary structure
f. concrete structure

1:10
Glass Wave Wall elevation/plan/section detail

the glass brick details combined as

Detail 7　细节图 7

Detail 8　细节图 8

SPACEPORT AMERICA
美国航天港

Architects: Foster + Partners
Client: New Mexico Spaceport Authority (NMSA)
Location: New Mexico, USA
Site Area: 300,000 ft^2
Photographer: Nigel Young

设计机构：Foster+Partners
客户：美国新墨西哥州航空总局
项目地址：美国新墨西哥州
占地面积：300 000 平方英尺
摄影：Nigel Young

Located in the desert-like landscape of New Mexico, Spaceport America is the first building of its kind in the world. Its design aims to articulate the thrill of space travel for the first space tourists while making a minimal impact on the environment. Viewed from space, the terminal evokes Virgin Galactic's brand logo of the eye, and is suggestive of an elongated pupil, with the apron completing the iris.

Organised into a highly efficient and rational plan, Spaceport has been designed to relate to the dimensions of the spacecraft. There is also a careful balance between accessibility and privacy. The astronauts' areas and visitor spaces are fully integrated with the rest of the building, while the more sensitive zones — such as the control room — are visible, but have limited access. Visitors and astronauts enter the building via a deep channel cut into the landscape. The retaining walls form an exhibition space that documents a history of space exploration alongside the story of the region and its settlers. The strong linear axis of the channel continues into the building on a galleried level to the super hangar — which houses the spacecraft and the simulation — through to the terminal building. A glazed facade on to the runway establishes a platform within the terminal building for coveted views out to arriving and departing spacecraft.

With minimal embodied carbon and few additional energy requirements, the scheme has been designed to achieve the prestigious LEED Gold accreditation. The low-lying form is dug into the landscape to exploit the thermal mass, which buffers the building from the extremes of the New Mexico climate as well as catching the westerly winds for ventilation and maximum use is made of daylight via skylights. Built using local materials and regional construction techniques, it aims to be both sustainable and sensitive to its surroundings.

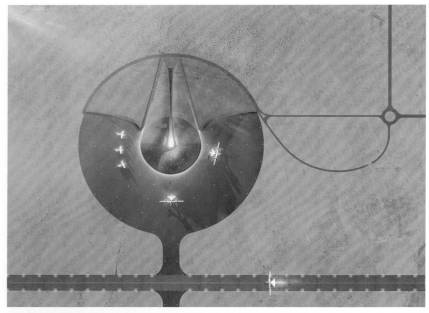

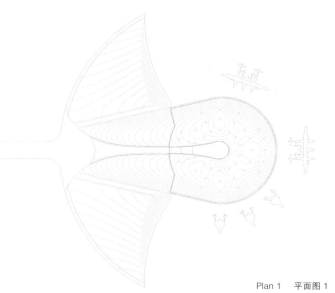

Plan 1　平面图 1

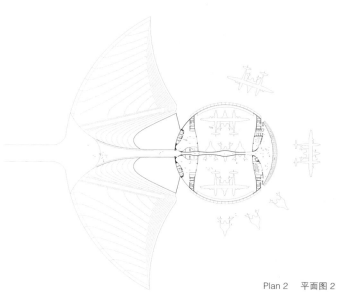

Plan 2　平面图 2

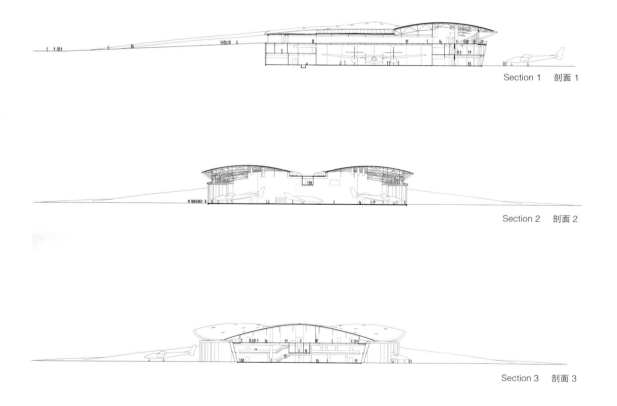

Section 1　剖面 1

Section 2　剖面 2

Section 3　剖面 3

Section 4　剖面 4

　　美国航天港位于新墨西哥州广袤的沙漠上，是世界上第一个商用太空机场。它的设计宗旨是激发第一次体验太空之旅的旅客的快感，同时把对环境的影响降到最低。从空间来看，此机场使人联想起维珍银河公司的标志——眼睛以及细长的瞳孔，整个停机坪如同一朵鸢尾花。

　　根据高效和合理的设计要求，此航天港大小的设计参考了宇宙飞船的实际大小。公共访问区和非开放区的设置很平衡。宇航员区域和访客空间与其余的建筑充分地融合在一起，而相当敏感的区域——如控制室——是可见的但具有很有限的访问权。游客和宇航员通过很深的地下通道进入此航天港。保留下来的墙壁形成了一个展览空间，记载着当地政府及当地居民探索太空的事迹。通道坚固的线型轴线一直延伸到航天港，走过一条长廊来到了飞机棚，飞机棚内有航天飞机和模拟室，而后就到达了航天港的终端。跑道旁的玻璃外墙建立起了一个平台，在终端大楼内可以一睹梦寐以求的航天飞机到达和离去的全过程。

　　此设计方案体现了最低排量的二氧化碳与极少额外的能源消耗需求，旨在得到享有盛名的 LEED 金牌认证。地势低洼是因为要到地下去探寻蓄热体，以此减缓新墨西哥州极端气候的影响，利用西风来通风以及最大限度地通过天窗采集自然光。同时结合使用当地材料和区域施工技术，设计旨在实现可持续发展和减少对环境的影响为目的。

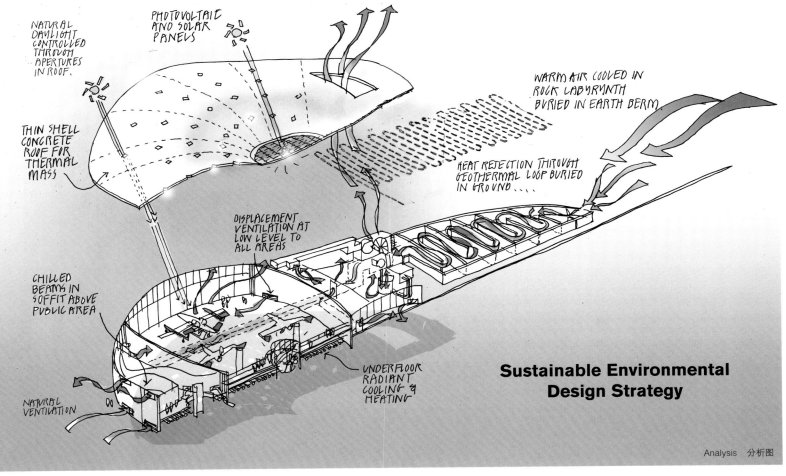

NATURAL DAYLIGHT CONTROLLED THROUGH APERTURES IN ROOF.

PHOTOVOLTAIC AND SOLAR PANELS

WARM AIR COOLED IN ROCK LABYRINTH BURIED IN EARTH BERM.

THIN SHELL CONCRETE ROOF FOR THERMAL MASS

HEAT REJECTION THROUGH GEOTHERMAL LOOP BURIED IN GROUND....

DISPLACEMENT VENTILATION AT LOW LEVEL TO ALL AREAS

CHILLED BEAMS IN SOFFIT ABOVE PUBLIC AREA

UNDERFLOOR RADIANT COOLING & HEATING

NATURAL VENTILATION

Sustainable Environmental Design Strategy

Analysis 分析图

TIANJIN GRAND THEATRE
天津大剧院

Architects: GMP · von Gerkan, Marg and Partners
Client: Tianjin Culture Centre Project and Construction Head Office
Project Leader: David Schenke, Xu Shan
Location: Tianjin, China
Building Area: 70,000 m²
Photographer: Christian Gahl

设计机构：GMP · von Gerkan, Marg and Partners
客户：天津文化中心项目建设办公室
项目经理：David Schenke, Xu Shan
项目地址：中国天津
建筑面积：70 000 平方米
摄影：Christian Gahl

Designers: Meinhard von Gerkan
and Stephan Schütz with Nicolas Pomränke
Chinese Partner practice: ECADI
Structural Engineers: schlaich bergermann und partner

设计师：Meinhard von Gerkan
and Stephan Schütz with Nicolas Pomränke
中国合作伙伴：ECADI
结构工程师：schlaich bergermann und partner

Tianjin Grand Theater occupies the key position in the newly built Culture Park of Tianjin. The circular shape of the roof construction corresponds with the existing Museum of Natural History so that an architectural dialogue of an earth-bound and a "floating" circular volume is created to both ends of the park. The earth and the sky represent a fundamental thinking in Chinese philosophy.

The roof volume of the Grand Theater opens up toward the broad water surface like an open sea shell. Opera hall, concert hall and the small multifunctional hall are exposed to the water surface like pearls inside this shell.

The three venues are conceived as free standing volumes on a stone base. Broad stairways connect the stone base with the raised plaza creating a kind of stage for urban life where people can overlooks the lake and the Culture Park.

Vehicular traffic is avoided on the water side. Drop-offs are located to the north and south whilst bus stops are situated along a dwelling mound to the east of the building. All internal areas are located within the base so that an unobstructed internal circulation is achieved.

The roof construction is a transformation of the traditional Chinese element of multiple eaves and thus defines a system of common horizontal layers, which creates an architectural entity of roof, facade and stone base.

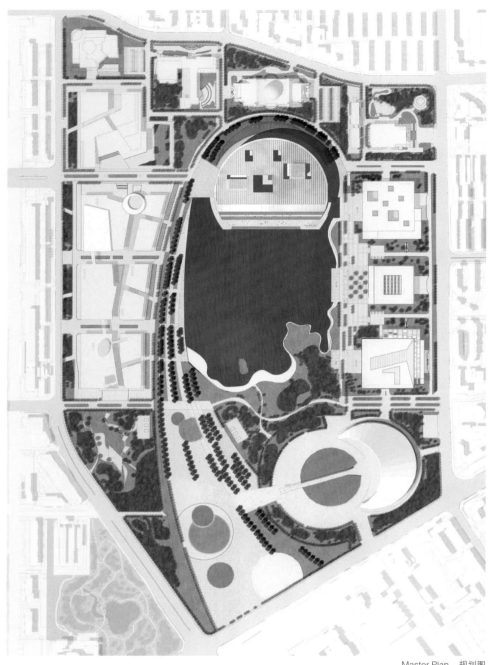

天津大剧院占据新落成的天津文化公园的核心位置。碟形的屋面结构与现有的自然历史博物馆相一致，在地面上的和"浮动"着的圆形体量之间形成建筑对话。天（天津大剧院）和地（自然历史博物馆）是中国哲学中的一对重要元素。

大剧院屋盖的半圆形体量朝向宽阔的水面打开，宛如张开的贝壳。剧院大厅、音乐厅和多功能厅如同贝壳中璀璨的珍珠，临水而立。

三座相互独立的演出大厅坐落于一个石质基座上，宽阔的台阶将石质基座和高处的平台相连，构成一个上演城市生活情景的舞台。人们可在此休憩，同时欣赏湖景，一览文化公园的全貌。

驶入区位于基地的南侧和北侧，车站位于东部建筑背向湖面的一侧，由此一来避免了在沿湖一侧设置机动车道。内部功能空间均统一置于建筑基座内，从而实现了高效率的内部交通流线。

屋面结构以富有时代感的设计表达了传统中式重檐庑殿顶的意象，并借鉴了中国古典建筑的三段式手法，通过横向的层次肌理，将屋面、立面、石基合为一体。

Master Plan　规划图

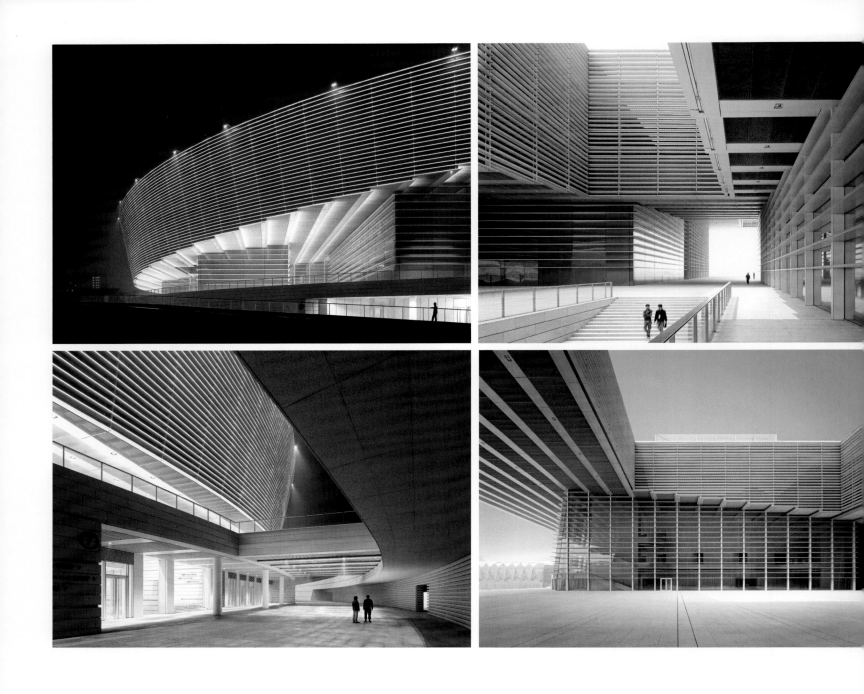

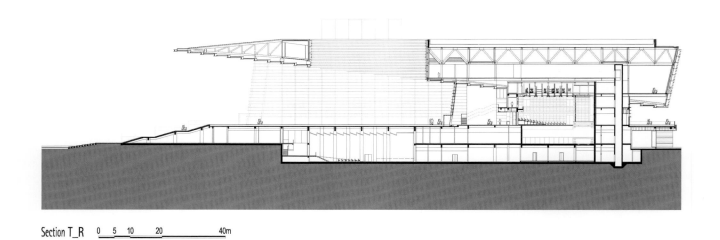

Section T_R   0 5 10   20   40m

Section 1　剖面图 1

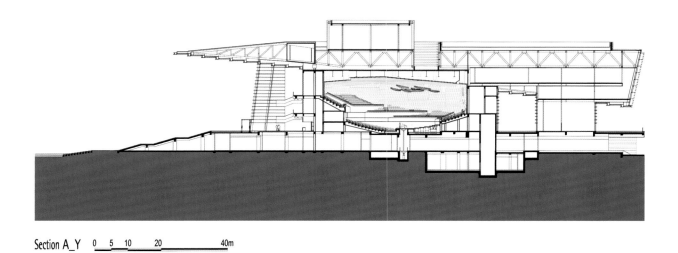

Section A_Y   0   5   10   20   40m

Section 2   剖面图 2

HARPA-REYKJAVIK CONCERT HALL AND CONFERENCE CENTRE
哈帕音乐会议中心

Architects: Henning Larsen Architects and Batteriid Architects
Client: Austurnhofn TR – East Harbour Project Ltd.
Landscape Architects: Landslag efh
Location: Reykjavik, Iceland
Area: 28,000 m²
Photographer: Nic Lehoux

设计机构：Henning Larsen Architects and Batteriid Architects
客户：Austurnhofn TR – East Harbour Project Ltd.
景观设计：Landslag efh
项目地址：冰岛雷克雅未克
面积：28 000 平方米
摄影：Nic Lehoux

Situated on the border between land and sea, the Concert Hall stands out as a large, radiant sculpture reflecting both sky and harbour space as well as the vibrant life of the city. The spectacular facades have been designed in close collaboration between Henning Larsen Architects, the Danish-Icelandic artist Olafur Eliasson and the engineering companies Rambøll and ArtEngineering Gmbh from Germany.

The Concert Hall of 28,000 m² is situated in a solitary spot with a clear view of the enormous sea and the mountains surrounding Reykjavik. The building features an arrival and foyer area in the front of the building, four halls in the middle and a backstage area with offices, administration, rehearsal hall and changing room in the back of the building. The three large halls are placed next to each other with public access on the south side and backstage access from the north. The fourth floor is a multifunctional hall with room for more intimate shows and banquets.

Seen from the foyer, the halls form a mountain-like mass as basalt rock on the coast forming a stark contrast to the expressive and open facade. At the core of the rock, the largest hall of the building is the main concert hall, which reveals its interior as a red-hot centre of force. The project is designed in collaboration with the local architectural company, Batteríið Architects.

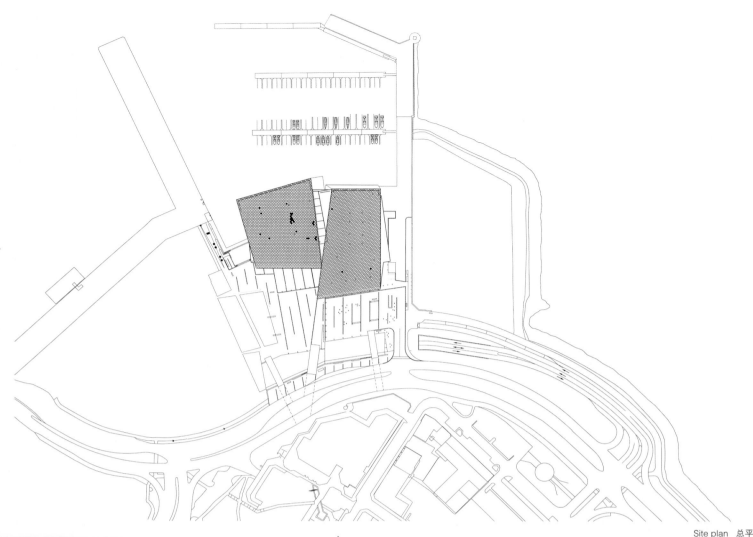

　　建筑坐落于陆地和海洋的边界上，音乐厅巍然屹立，像一尊巨大的雕塑辐射四方，映衬天空和海港以及这座城市充满活力的生活。其出色的外观是在 Henning Larsen 建筑事务所、丹麦裔冰岛艺术家 Olafur Eliasson、Rambøll 工程公司以及来自德国的 ArtEngineering GmbH 的密切合作下设计的。

　　面积达 28 000 平方米的音乐厅位于一个僻静之处，视野开阔，能看到辽阔的大海和雷克雅未克周围的群山。该建筑的前部设有一个大堂区、中部四厅和一个带办公室的后台区，行政区、彩排间和更衣室则在后部。三间大厅彼此毗邻，以南侧的公众通道和北边的后台通道相连。第四层是一个多功能厅，里面有为演出和宴会而设的房间。

　　从出入口大堂看去，所有的厅构成一个山状的地块，与海岸上的玄武岩形成鲜明的比照。在岩石的内核中，最大的厅即音乐主厅，是其内部力量的炽热中心。该项目是在本地建筑公司 Batteríið 事务所的协助下设计的。

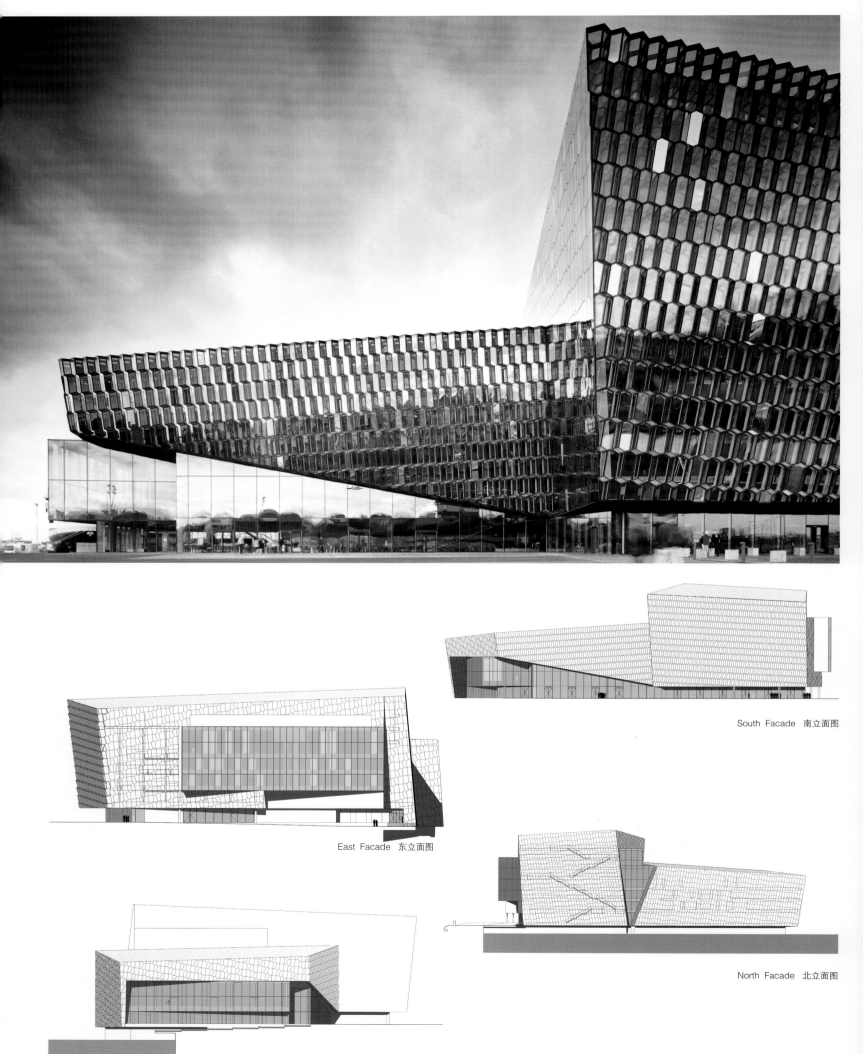

East Facade 东立面图

South Facade 南立面图

West Facade 西立面图

North Facade 北立面图

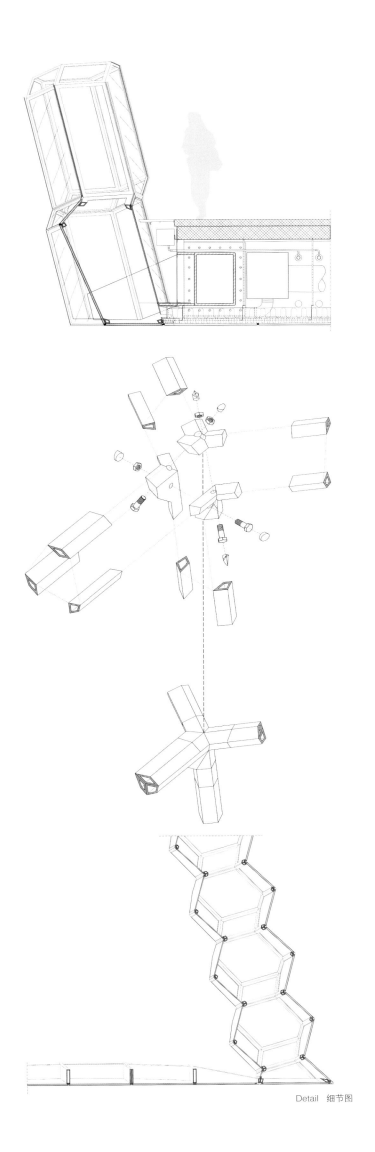

Detail 细节图

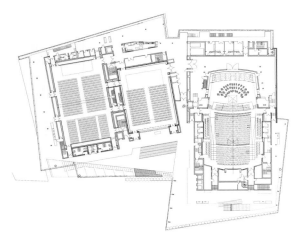

Level 2 Plan　二层平面图

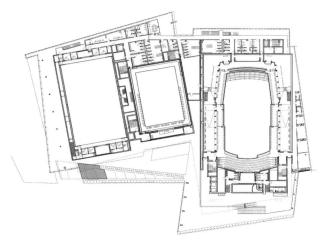

Level 3 Plan　三层平面图

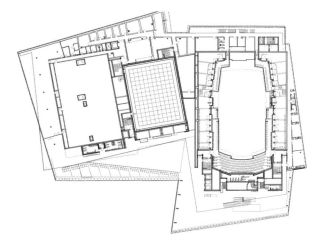

Level 4 Plan　四层平面图

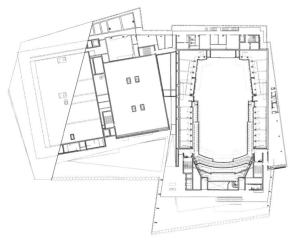

Level 5 Plan　五层平面图

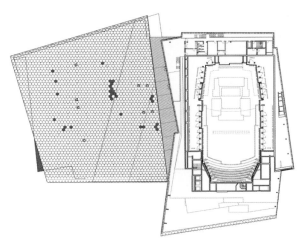

Level 6 Plan　六层平面图

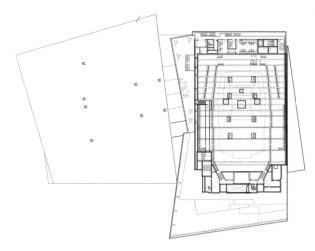

Level 7 Plan　七层平面图

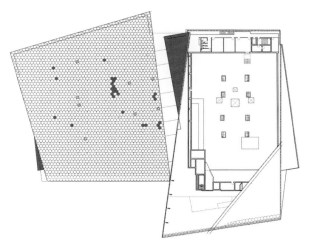

Level 8 Plan　八层平面图

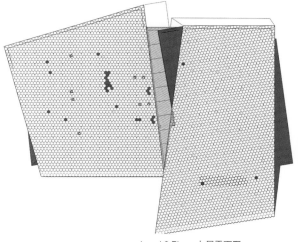

Level 9 Plan　九层平面图

Level 1 Plan 一层平面图

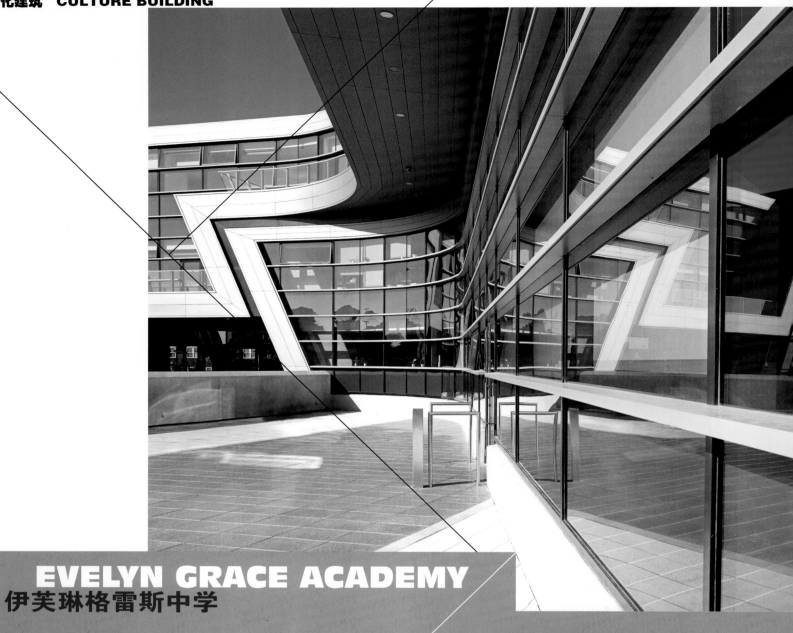

EVELYN GRACE ACADEMY
伊芙琳格雷斯中学

Architects: Zaha Hadid Architects
Client: School trust: ARK Education Government: DCSF
Location: London, UK
Area: 10,745 m^2
Photographer: Hufton + Crow

设计机构：Zaha Hadid Architects
客户：School trust: ARK Education Government: DCSF
项目地址：英国伦敦
面积：10 745 平方米
摄影：Hufton + Crow

The Evelyn Grace Academy in Brixton, London Borough of Lambeth, offers an opportunity to broaden not only the educational diversity of this active and historic part of London, but also to augment the built environment in a predominant residential area. This Academy presents itself as an open, transparent and welcoming addition to the community's local urban regeneration process.The strategic location of the site within two main residential arteries naturally lends the built form to be coherent in formation and assume a strong urban character and identity, legible to both the local and neighbouring zones.

The Academy offers a learning environment that is spatially reassuring and able to engage the students actively, creating an atmosphere for progressive teaching.To maintain the educational principle of schools-within-schools, the design generates natural patterns of division within highly functional spaces which give each of the four smaller schools a distinct identity, both internally and externally. These spaces present generous environments with maximum levels of natural light, ventilation and understated but durable textures. The communal spaces — shared by all the schools — are planned to encourage social communication with aggregation nodes that weave together the extensive accommodation schedule.

Similarly, the external shared spaces, in order to generate a setting that encourages interaction, are treated in a manner of layering to create informal social and teaching spaces at various levels based on the convergence of multiple functions. The vision of the scheme provides an educational complex that will be equally esteemed and cherished by the pupils and community.

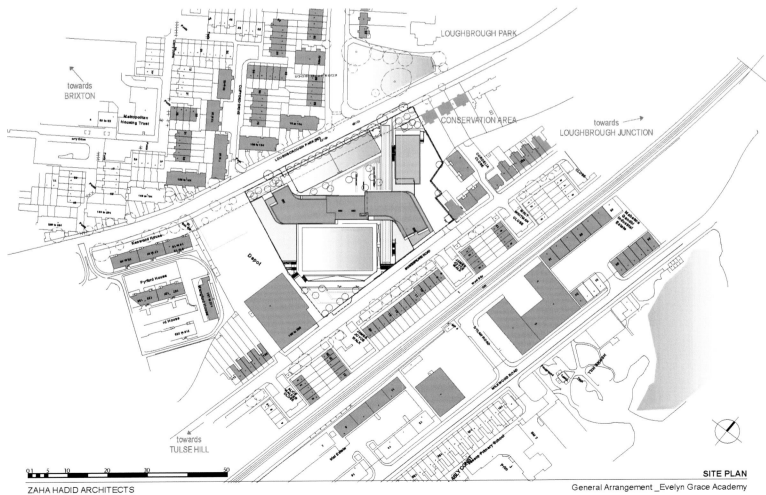

towards
BRIXTON

LOUGHBROUGH PARK

CONSERVATION AREA

towards
LOUGHBROUGH JUNCTION

towards
TULSE HILL

SITE PLAN

General Arrangement _Evelyn Grace Academy

Plan 1　平面图 1

　　伊芙琳格雷斯中学位于伦敦兰贝斯区的布里克斯顿。该中学不仅为这个活跃且历史悠久的地区增添了教育的多样性，同时也改善了这一住宅区域占主导地位的建筑环境，并在该社区的都市重建工程中扮演了一个开放、透明而且好客的附加项目的角色。该中学位于两条住宅区主干道之间的有利位置，使建筑形式从结构上得到了统一，不管是在当地还是邻近区域，都能领略到其强烈的城市特征。

　　建筑提供了安静的学习环境，使学生们更加积极地参与到学习中，创造了不断进步的教学环境。为了遵循"校中校"的教学原则，大楼的设计在高度功能化的空间中创造出自然的分隔形式，从内部到外部同时赋予四个小型学校不同的特性。这些空间最大化地享有自然采光和通风，质朴的材料围合出良好的环境。共用空间的设计促进了校园内的社交活动，这些区域可举行参与人数众多的活动。

　　同样，为了创造这样一个互动的环境，外部共享空间采用分层的形式，在不同楼层分设社交空间和教学空间。项目规划为学生和社区提供了一个珍贵的和值得尊重的教育综合体。

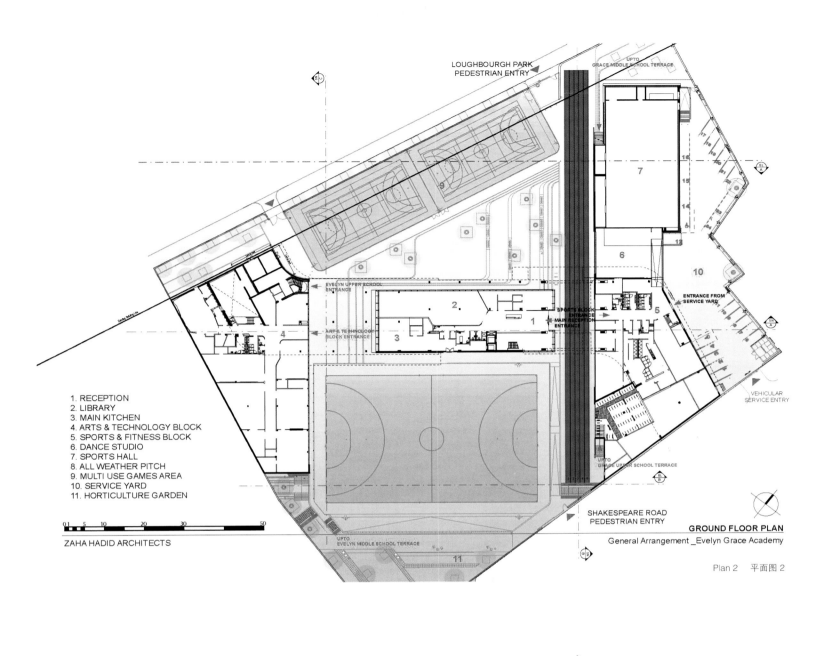

1. RECEPTION
2. LIBRARY
3. MAIN KITCHEN
4. ARTS & TECHNOLOGY BLOCK
5. SPORTS & FITNESS BLOCK
6. DANCE STUDIO
7. SPORTS HALL
8. ALL WEATHER PITCH
9. MULTI USE GAMES AREA
10. SERVICE YARD
11. HORTICULTURE GARDEN

ZAHA HADID ARCHITECTS

GROUND FLOOR PLAN
General Arrangement _Evelyn Grace Academy

Plan 2   平面图 2

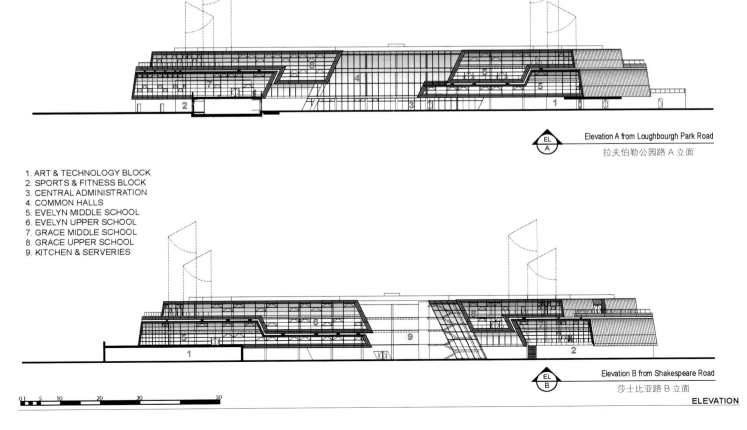

1. ART & TECHNOLOGY BLOCK
2. SPORTS & FITNESS BLOCK
3. CENTRAL ADMINISTRATION
4. COMMON HALLS
5. EVELYN MIDDLE SCHOOL
6. EVELYN UPPER SCHOOL
7. GRACE MIDDLE SCHOOL
8. GRACE UPPER SCHOOL
9. KITCHEN & SERVERIES

Elevation A from Loughbourgh Park Road
拉夫伯勒公园路 A 立面

Elevation B from Shakespeare Road
莎士比亚路 B 立面

ELEVATION

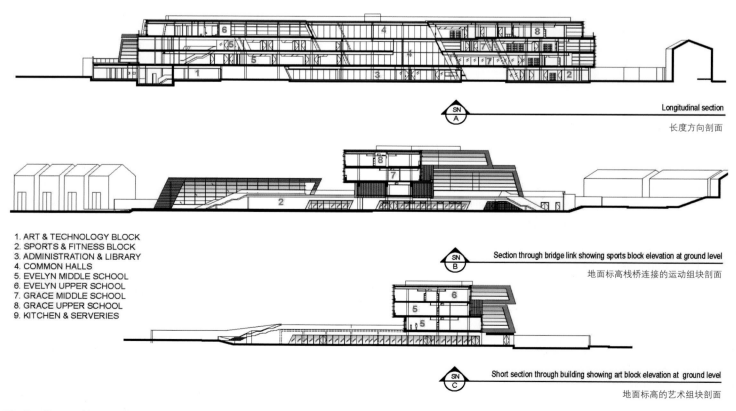

1. ART & TECHNOLOGY BLOCK
2. SPORTS & FITNESS BLOCK
3. ADMINISTRATION & LIBRARY
4. COMMON HALLS
5. EVELYN MIDDLE SCHOOL
6. EVELYN UPPER SCHOOL
7. GRACE MIDDLE SCHOOL
8. GRACE UPPER SCHOOL
9. KITCHEN & SERVERIES

Longitudinal section

长度方向剖面

Section through bridge link showing sports block elevation at ground level

地面标高栈桥连接的运动组块剖面

Short section through building showing art block elevation at ground level

地面标高的艺术组块剖面

0 1 5 10   20   30   50

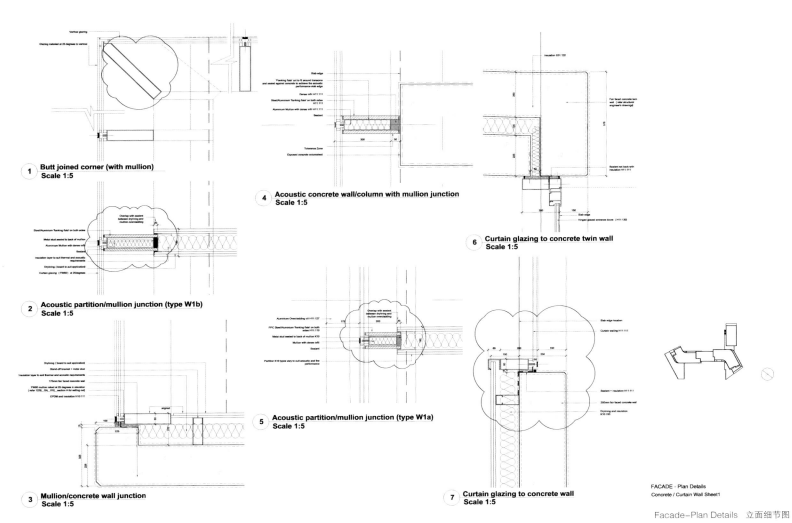

1 Butt joined corner (with mullion)
Scale 1:5

2 Acoustic partition/mullion junction (type W1b)
Scale 1:5

3 Mullion/concrete wall junction
Scale 1:5

4 Acoustic concrete wall/column with mullion junction
Scale 1:5

5 Acoustic partition/mullion junction (type W1a)
Scale 1:5

6 Curtain glazing to concrete twin wall
Scale 1:5

7 Curtain glazing to concrete wall
Scale 1:5

FACADE - Plan Details
Concrete / Curtain Wall Sheet1

Facade–Plan Details 立面细节图

FLUSHING MEADOWS-CORONA PARK NATATORIUM & ICE RINK
法拉盛 MEADOWS–CORONA
公园游泳池及滑冰场

Architects: Handelarchitects
Designer: Blake Middleton, FAIA
Location: New York City, USA
Building Area: 10,219m^2
Photographer: David Sundberg

设计机构：美国汉德建筑
设计师：Blake Middleton, FAIA
项目地址：美国纽约
建筑面积：10 219 平方米
摄影：David Sundberg

The Flushing Meadows-Corona Park Natatorium & Ice Rink is situated on the edge of Flushing Meadows-Corona Park and next to the Van Wyck Expressway, where it mediates between the urban neighborhood and open green space. It is recognized as the largest recreation facility ever built in a New York City Park and the first indoor public pool to be built in four decades.

The City of New York required a new facility for the growing population of northern Queens. The rink was to replace an aging facility in the Queens Museum. Construction of a previous design started but was halted by funding constraints after September 11, 2001. When the city restarted the project, the pool became part of New York City's 2012 Olympic bid, and had to meet requirements for an Olympic water polo venue. To save time and money, the previously installed piles were retained. The project had to meet a fast track delivery schedule, and cost no more than $67 million.

The facility includes an Olympic-sized indoor swimming pool with seating for 400, a NHL regulation-size ice rink with seating for 440, locker rooms, and a lobby. The swimming pool is ADA accessible; one-third of the pool floor moves vertically and also includes two moveable bulkheads to configure the swimming area for different competitions. The pool and rink are oriented end-to-end, juxtaposing liquid and frozen water, humid and dry environments. An outdoor terrace is accessible from the pool deck.

In the dynamic spirit of the 1939 and 1964 World's Fair pavilions that occupied the site, the canopy-like roof is suspended over the natatorium and rink, punctuated with soaring masts supporting a cable stay structure. The entire envelope is composed of high performance precast concrete panels and a curtain wall opening out from the pool to dramatic park views.

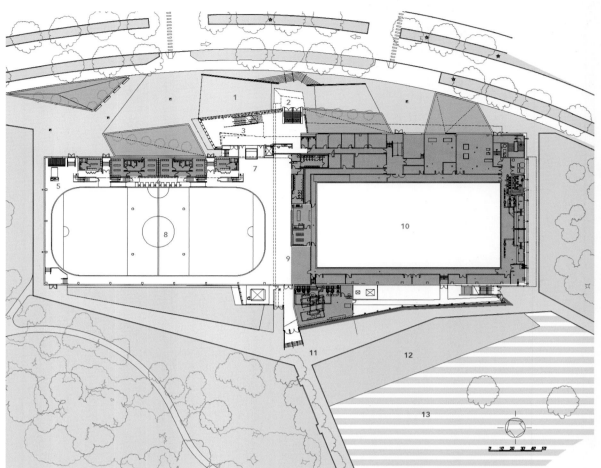

Ground Floor Plan

1   North Entry Plaza
2   North Entry
3   Lobby
4   Offices
5   Zamboni
6   Locker Rooms
7   Ice Rink Entry
8   Ice Rink
9   Concession
10   Pool Vessel
11   Park Entry
12   Green Beach
13   South Meadow

Ground Floor Plan　底层平面图

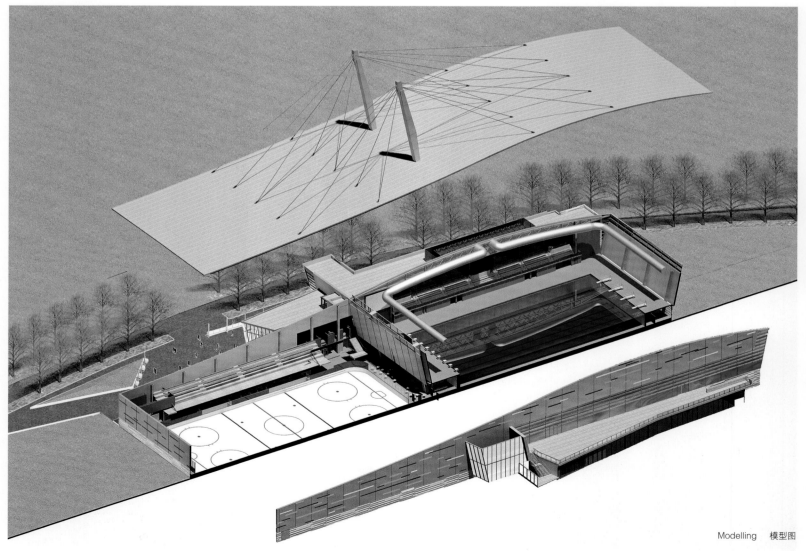

Modelling　模型图

法拉盛 Meadows-Corona 公园游泳池及滑冰场位于法拉盛 Meadows-Corona 公园边缘,紧邻 Van Wyck 高速公路,是城市社区和开放绿地的过渡空间。这是纽约城市公园最大的娱乐设施,也是 40 年来首个室内公共游泳馆。

纽约北部皇后区日益增长的人口需要新的公共设施,滑冰场取代了皇后区博物馆的老旧设施。项目动工之后,由于经费紧张,在 2001 年 9 月 11 日停止施工。当项目重新开始时,游泳池成为了纽约申办 2012 年奥林匹克运动会竞标场馆,因此需要满足奥运会对水球场地的要求。为了省时省钱,建筑仍保留原来的墩桩。项目建设工期紧张,花费不超过 6 700 万美元。

建筑包括一个可容纳 400 人的奥林匹克规格的游泳池、一个可容纳 440 人的 NHL 规格的滑冰场以及更衣室和大厅。游泳池可以无障碍通行,三分之一的游泳池楼层可以垂直升降,两块可移动的隔墙可以随意布局游泳区域以应对不同的赛事。游泳池和滑冰场均是头尾连接,水和冰、湿润和干燥的环境互不干扰。游泳池内平台与室外平台连通。

作为 1939 年和 1964 年的世博会所在地,华盖般的屋顶横跨游泳池和滑冰场,高耸的桅杆结构支持着长跨距的滑冰场和游泳池。建筑外壳由高性能的混凝土板和开放式的幕墙组成,可让游客享受室内外视野。

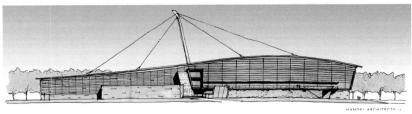

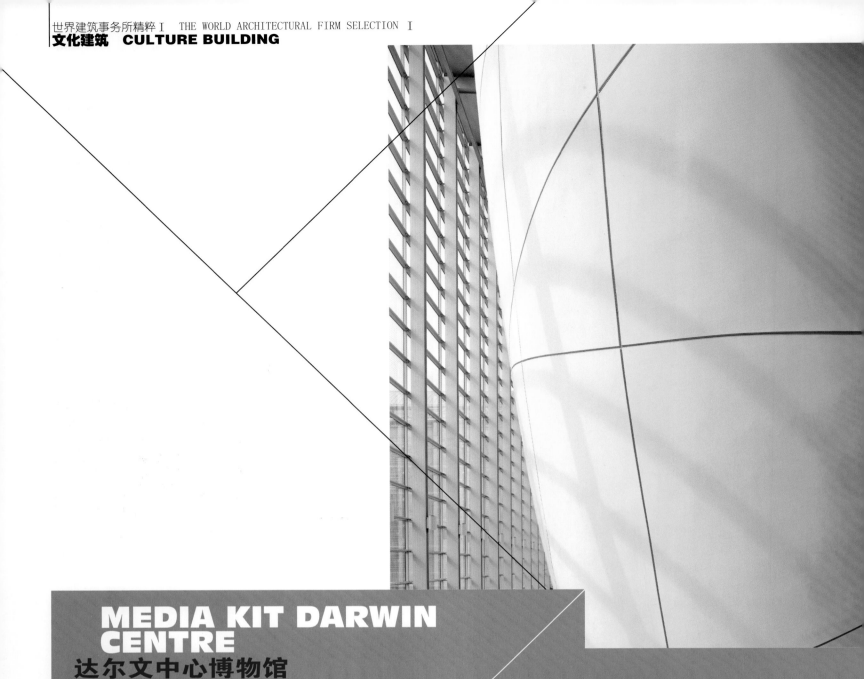

MEDIA KIT DARWIN CENTRE
达尔文中心博物馆

Architects: C.F Møller_Architects
Partner and Architect: Anna Maria Indrio
Client: Natural History Museum
Project Manager: Manly Development Services
Structural Engineer: Arup
Location: Cromwell Road, London, SW7
Area: 16,000 m²

设计机构：C.F Møller_Architects;
主设计师及合作人：Anna Maria Indrio
客户：Natural History Museum
项目经理：Manly Development Services
结构工程师：Arup
项目地址：伦敦西南七区克伦威尔路
面积：16 000 平方米

Visualising the collection
Expressing a unique and appropriate architectural concept.
The design of the second phase of the Darwin Centre project is characterised by a compelling and strong architectural concept in order to contain and represent vast entomological and botanical collections housed within the Natural History Museum.

The cocoon
The solution to resolving the various client requirements and to clearly symbolize the world class collection of specimens is the "Cocoon" — an architectural translation which forms the inner protective envelope.
The scale of the Cocoon form is such that it cannot be seen in its entirety from any one position. This emphasizes its massive scale. The shape and size give the visitor a tangible understanding of the volume of the collections contained within.
The collections housed in the Natural History Museum are among the world most extensive and treasured. In order to adequately preserve, maintain and represent this collection, a structure suitable in both its expression and physical construction is necessary. The Cocoon does this by creating an icon, which represents preservation, protection and nature. It is constructed of 300 mm thick walls, with a defined geometric form based on mathematical equations. The surface finish is ivory-coloured polished plaster, resembling a silk cocoon, in which a series of expansion joints wrap around, resembling silk threads.

Respecting the site
The second phase of the Darwin Centre is intended to manage the difference in scale, architectural approach and to create a physical link between the original landmark Alfred Waterhouse Museum building and the more contemporary addition of the first phase of the Darwin Centre. It also serves as a landmark building in its own right, the full height glass wall partially revealing the solid 3-dimensional form of the cocoon within.

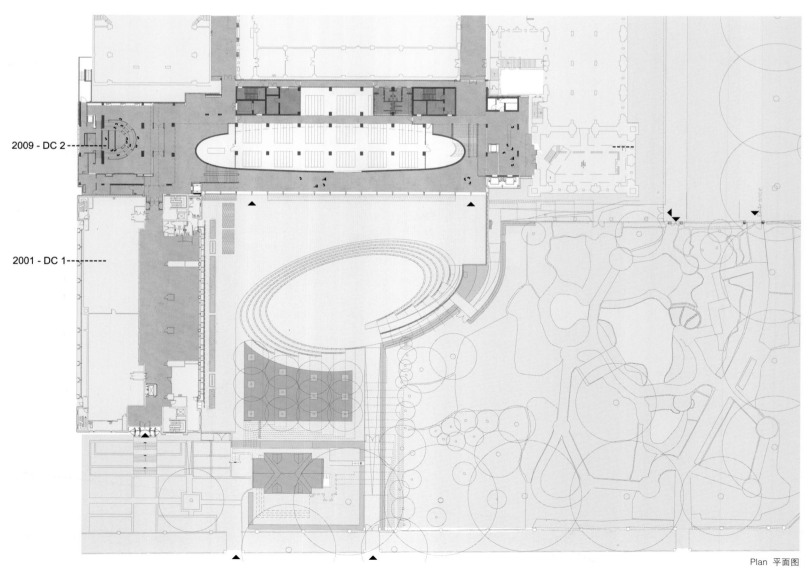

2009 - DC 2----

2001 - DC 1----

Plan 平面图

Bridging past, present and future
The second phase of the Darwin Centre improves and transforms the existing buildings into something more than the sum of its parts. The new building links existing and new buildings into a dialogue forming a set of dynamic, spatial experiences, bridging the past, present and future for the museum.
The smooth curved form of the immense cocoon is an iconic feature of the new Darwin Centre building and the public atrium space is dramatic, tall and filled with daylight.

Maximizing access for all
Public access to the scientific core of the second phase of the Darwin Centre takes the form of a visitor route up and through the cocoon, overlooking the science and collection areas without compromising the central activities of protection, preservation and research.
Passing through the Cocoon, the visitor enters a new space where the boundaries between the inner and outer worlds of scientific research are blurred.
The visitor can experience the Darwin Centre as a compelling and interactive learning space, observing the scientific and research activities without interrupting scientific work in progress.

Section 剖面图

展览藏品

传达一种独特而相称的建筑学理念。

达尔文中心项目第二期的设计，由引人注目的强烈的建筑理念定调，试图借此蕴含和代表安置在自然史博物馆内的大量昆虫和植物收藏品。

"蚕蛹"

"蚕蛹"经由建筑转化构成内部的保护壳层，是解决客户各种要求的方法，也代表着世界级水平的标本收藏。

"蚕蛹"规模庞大，以至于无论你身处里面哪个位置，都无法看全它的整体。如此一来，更显其辐射之广。它的形状和大小让游客对里面的收藏品有一种触手可及的感觉。

博物馆中的收藏品是全世界种类最齐全的，也是最珍贵的。为了完好地保存、维护和展示这些收藏，需要一个表面气质和内在结构相得益彰的结构体。以形名状的"蚕蛹"即是如此的结构体，它创造了一个代表防腐、保护和自然的圣像。它由30厘米厚的墙建成，有着基于数学公式的几何形状。建筑表面以象牙色打磨石膏抛光，形如绢蚕，周遭包裹着如同丝线的伸缩缝。

有关馆址

达尔文中心第二期要达到在规模和建筑方式上造成差异的目的，同时建立起与原来的地标阿尔弗雷德·沃特豪斯博物馆和第一期更为现代化的附属楼之间的物理学联系。在其右侧，它也会成为一座地标，而全高玻璃墙部分展露出茧体内部坚实的三维构架。

沟通过去、现在和未来

达尔文中心第二期将现今的建筑群提升和转化成某种大于其所有部分总和的东西。新建筑把已有建筑和新楼群引入一场对话，构成一对充满活力和空间经验的组合，沟通了博物馆的过去、现在和将来。

巨型茧体流畅的弯曲形式是新达尔文中心的标志性象征，其中的公共体量最大化地沟通了天井，使其日光盈溢，显得宽阔高远。

去往达尔文中心第二期的科技区的公众通道，采用游客从上穿过茧体的形式，能够俯瞰科技区和收藏区而不妨碍保养、存放和研究等重要活动。

穿过茧体，游客进入一个崭新的空间，在那里科学研究的内外世界之间的界限模糊不清。

游客可以把达尔文中心当作一个引人入胜的互动的学习空间来经历，观摩科学研究活动而不至于干扰其工作进度。

OLYMPIC TENNIS CENTRE
奥林匹克网球中心

Architects: Dominique Perrault Architecture
Client: Madrid Espaciosy Congresos
Project Manager: LKS
Location: Madrid, Spain
Building Area: 100,000 m²
Photographer: DPA/ADAGP

设计机构：Dominique Perrault Architecture
客户：Madrid Espaciosy Congresos
项目经理：LKS
项目地址：西班牙马德里
建筑面积：100 000 平方米
摄影：DPA/ADAGP

In the prospect of being a candidate for staging the Olympic Games in 2016, the Spanish capital has begun a campaign to build spectacular facilities among the Olympic Tennis Centre. Working in an indistinct peripheral area, the issue at stake was not so much to design a building as to stage to manage architecture and invent scenery.

The "magic box" concept encloses sports and multi-functional buildings but opens up and shapes itself to the various uses projecting a changing and lively silhouette in the cityscape. Its mobile and vibrant skin filters the sunlight, serves as a windbreak and shelters the sports halls in a ligh-aeight shell.

Water forms a lake to define a wide horizontal plane of reference, like a huge natural mirror. Islands invite visitors to enjoy the pleasure of calm walks or sports activities.

All tells an architectural landscape that flows and ripples like a garment, a place for strolling and having fun, a venue that is alive day and night.

为了申办 2016 年奥运会，西班牙首都马德里正在为奥林匹克网球中心修建设施，这个在边界模糊的区域进行的工程与其说是在设计一幢建筑，不如说是在创造一处风景。

"魔盒"的概念在将体育场和多功能建筑封闭的同时开放自身，使自己具备多种用途，在城市景观中形成多变且充满活力的剪影。那可移动且充满动感的轻质量表皮可以过滤阳光，并作为挡风墙来保护体育馆。

水形成了一个湖泊，像一面自然之镜定义了宽阔的水平面。湖中的岛屿邀请来访者享受悠闲散步或运动的愉悦。

这里的建筑景观就像衣服上的流线和褶皱。这里是一个漫步的好地方，也是一个日日夜夜都充满生机的体育场所。

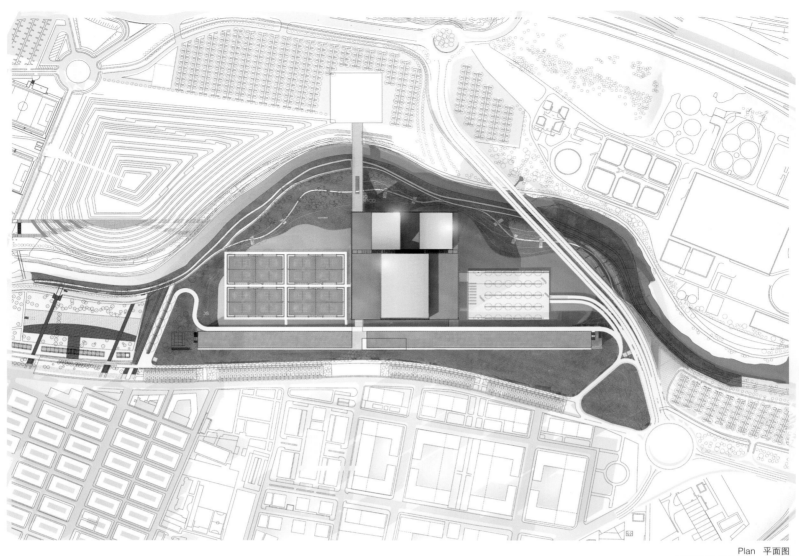

Plan 平面图

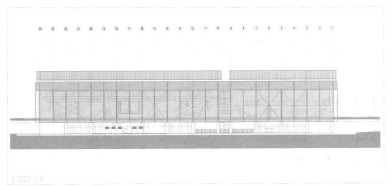

Elevation 1　立面图 1

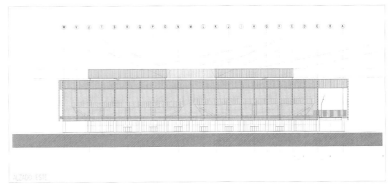

Elevation 2　立面图 2

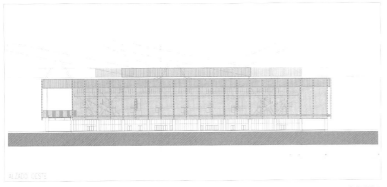

Elevation 3　立面图 3

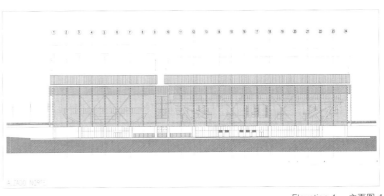

Elevation 4　立面图 4

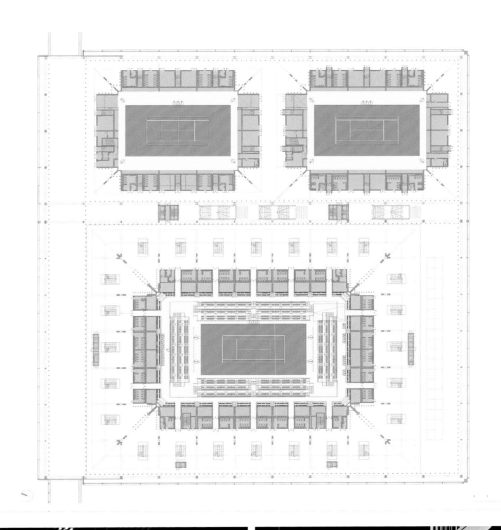

Plan 平面图

EWHA WOMANS UNIVERSITY
梨花女子大学

Architects: Dominique Perrault Architecture
Partners: Baum Architects
Client: Ewha Womans University
Location: Seoul, Korea
Building Area: 70,000 m²
Photographer: DPA/ADAGP

设计机构：多米尼克·佩罗建筑师事务所
合作者：Baum Architects
客户：韩国梨花女子大学
项目地址：韩国首尔
建筑面积：70 000 平方米
摄影：DPA/ADAGP

Ewha Womans University's new building (founded in 1886, Ewha welcomes 22,000 female students and is ranked as one of the best universities in the world), designed by Dominique Perrault Architecture, is a result of an international architecture competition organised in 2003, and inaugurated on April 29th 2008.

A landscape then, more than an architecture work, is located in the midst of Seoul's university area. The sport grounds, event locations and educational buildings mix, intermingle and follow one after another in campus valley. A long asphalted strip, delineated at one end of a race track, is completely surrounded by nature landscape of pear trees and shrubs. Black asphalt, red race track, green nature and finally the white brightness of the valley interweave here. A valley, which is bravely drawn in the ground, slides down along a gentle slope. At the other end of the valley, the slope becomes a huge stairway which can be used as an open air amphitheatre if necessary.

At the very heart of the valley, a dreamlike impression takes place. Opposed to the outdoor world, a subtle and serene universe appears suddenly. Classrooms and libraries, amphitheatres and auditoriums, shops and movement center... Everything enjoy the natural light.

Perrault is prone to buried, excavated, nestled places (the French National Library in Paris, the Velodrome and Olympic swimming pool in Berlin, the Kansai Library in Japan and the Cultural Centre in Santiago de Compostela...) Perrault has the desire, physically speaking, to appropriate the territory, to mingle the construction with the ground, the desire to exploit to its paroxysm the idea that "concept and matter have to grapple one with another".

At Ewha Womans University, Perrault put his heart in action: words (idea, concept, abstraction, geometry, strategy, tension, fusion, freedom, simplicity, evidence...), principles (physics, mechanics, dimension...) and commitments (urban concerns, creation of a location and not only of a building, refusal of formalism, and disappearance of architecture...) which best qualify his architecture.

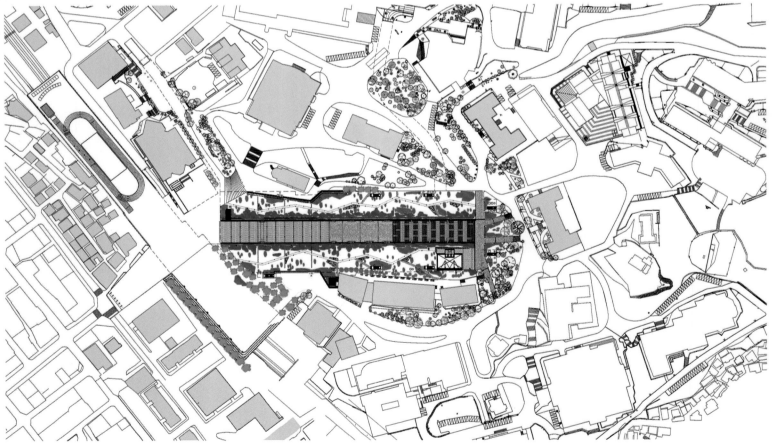

Plan 平面图

ROOF PLAN
0 10 20   50

242

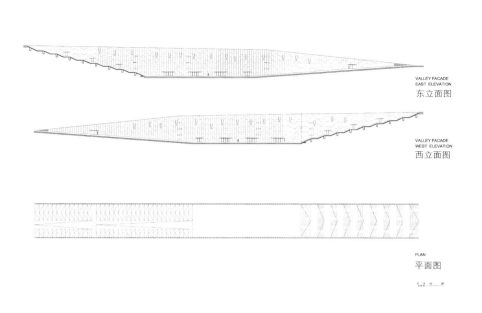

VALLEY FACADE
EAST ELEVATION
东立面图

VALLEY FACADE
WEST ELEVATION
西立面图

PLAN
平面图

EAST ENTRY ELEVATION
东入口立面图

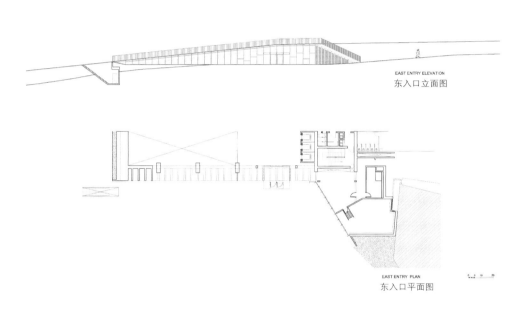

EAST ENTRY PLAN
东入口平面图

WEST ENTRY ELEVATION
西入口立面图

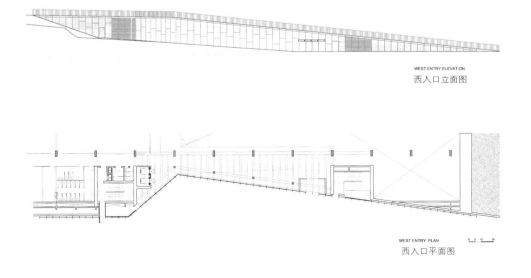

WEST ENTRY PLAN
西入口平面图

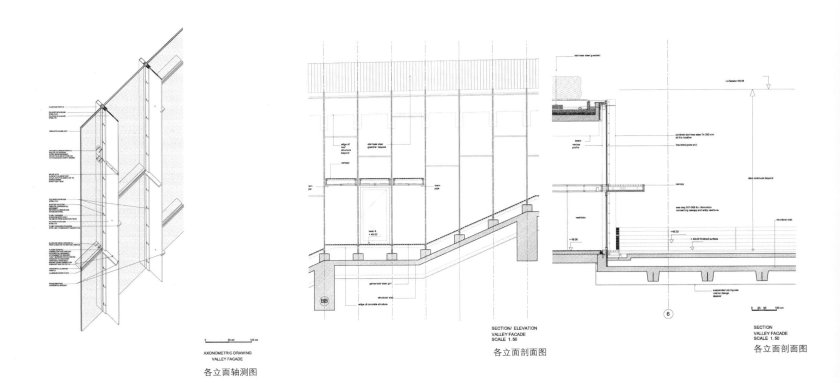

AXONOMETRIC DRAWING
VALLEY FACADE

各立面轴测图

SECTION/ ELEVATION
VALLEY FACADE
SCALE 1:50

各立面剖面图

SECTION
VALLEY FACADE
SCALE 1:50

各立面剖面图

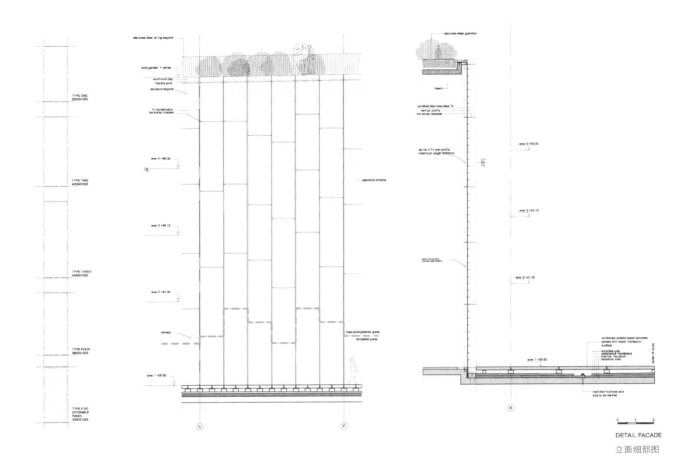

DETAIL FACADE
立面细部图

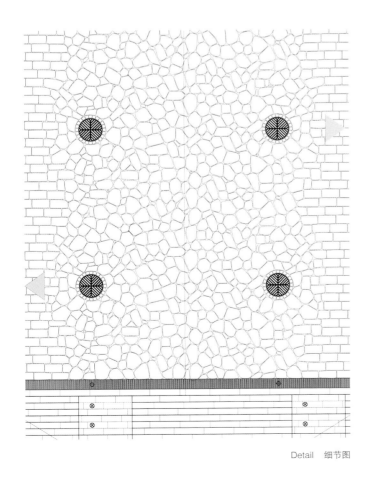

Detail 细节图

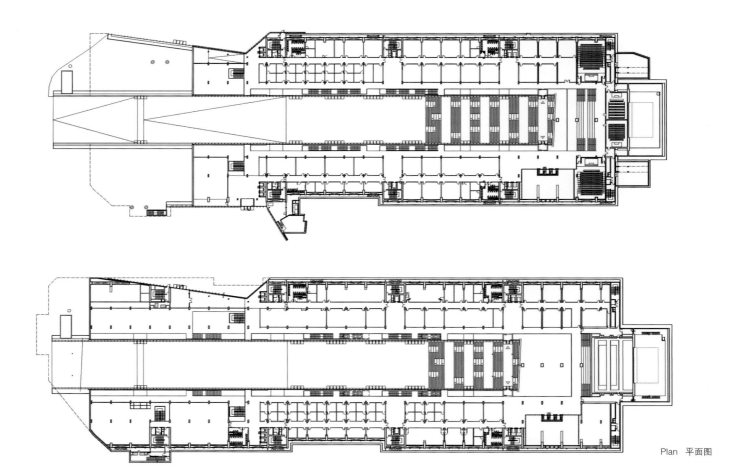

Plan 平面图

梨花大学新大楼由法国的 Dominique Perrault 建筑师事务所设计（梨花大学建于 1886 年，拥有 22 000 名女学生，是世界上最好的大学之一），此设计方案是 2003 年国际建筑竞赛的获奖作品，于 2009 年 4 月 29 日动工。

项目位于首尔中部大学区，它不仅仅是一个建筑，更是一个景观。校园里的运动场地、活动地点及教育建筑交织混合，跑道的一头有一条长长的柏油路，完全被梨树和灌木创造的自然景观所包围。黑色沥青、红色跑道、绿色的大自然和白色的灯火在这里交汇。山谷沿着舒缓的角度向下延伸，仿佛一幅地面上的华丽图画。在山谷的另一端，斜坡被设计成大型阶梯，需要时可以当作一个露天剧场。

山谷的中心给人一种朦胧的感觉。与外面的世界相比，这里是一个微妙而宁静的地方。教室和图书馆、露天剧场和礼堂、商店和运动中心……这里的一切都享受着自然之光。

Perrault 倾向于建造埋入式的、可开挖的和鸟巢状的场所（如法国巴黎的法国国家图书馆、柏林的赛车场和奥运游泳池、日本的关西图书馆和圣地亚哥德孔波斯特拉的文化中心等），佩罗希望适当利用建设用地，探索出一种爆发性的方式，使"概念和问题一个一个地解决掉"。

在梨花女子大学的设计中，Perrault 将自己的设计理念贯彻落实：语言（思想、概念、抽象概念、几何学、策略、张力、融合、自由、简单、明确……）、原则（物理学、力学、尺寸……）和承诺（城市问题、建设一个外景而不仅仅是一个建筑、拒绝形式主义、建筑风格消失……）均在他的建筑里得到了最好的体现。

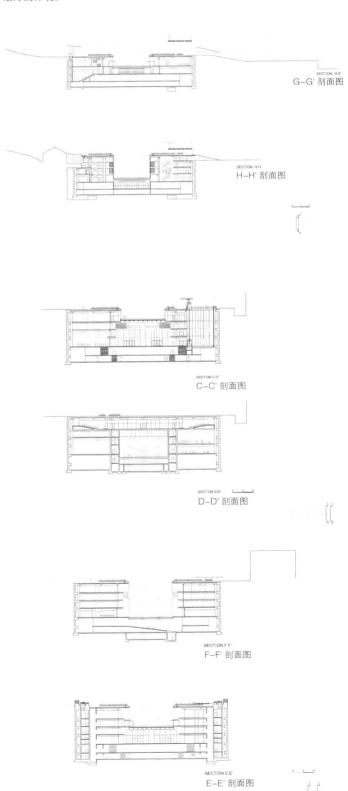

G-G' 剖面图

H-H' 剖面图

C-C' 剖面图

D-D' 剖面图

F-F' 剖面图

E-E' 剖面图

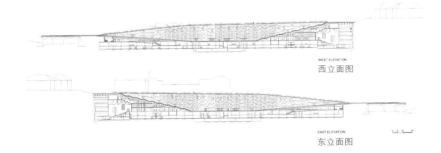

西立面图

东立面图

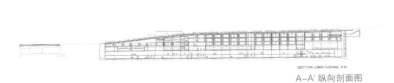

A-A' 纵向剖面图

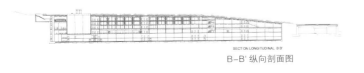

B-B' 纵向剖面图

FERRARI WORLD ABU DHABI
阿布扎比法拉利世界主题公园

Architects: Benoy
Location: Yas Island, Abu Dhabi
Area: 236,000 m²

设计机构：Benoy
项目地址：阿布扎比亚斯岛
面积：236 000 平方米

The world's first Ferrari Theme Park is situated at the centre of the landmark Yas Island mega scheme in Abu Dhabi, providing thrilling and multi-sensory experience for visitors.

Ferrari World Abu Dhabi's location, scale and purpose presented enormous architectural challenges. In response, Benoy delivered a revolutionary design solution. The end result is an iconic landmark leisure destination that reflects both the integrity of the Ferrari brand and the ambitions of Abu Dhabi.

Opened in December 2010, Ferrari World Abu Dhabi is the world's largest indoor theme park. Benoy's vision was to create a building that would reflect the highly recognizable sinuous form of Ferrari, directly inspired by the classic double curve profile of the Ferrari GT chassis.

The building was conceived as a simple "Ground hugging" structure: a red sand dune. The 3D nature of the building was derived from the sinuous double curve of the classic Ferrari body shell. The double curve was proportionately applied in elevation to set the structure's length (700 m) and height (45 m).

The metal skin roof is highly insulated and the main facades utilise efficient glass to reduce thermal loads and glare.

At the centre, the roof dips and gathers itself into the ground in the form of a crystal glazed lit funnel, creating the perfect location for one of the most exhilarating rides: the 60 m high "G-Force Tower".

Ferrari World Abu Dhabi houses more than 20 high octane attractions under a geometrically compelling space frame structure. With state-of-the-art simulators allowing guests to experience the thrill of racing a Ferrari and the world's fastest roller coaster reaching speeds in excess of 200 kph, Ferrari World Abu Dhabi is an awe-inspiring experience.

世界上第一个"法拉利世界"主题公园坐落在阿布扎比的地标——亚斯岛规划区，其内众多颇具创造性的游乐设施将给游客带来前所未有的多方位感官体验。

阿布扎比法拉利主题公园的位置、规模和目标面临大量的设计难题，Benoy采用具有革命性的设计方案，最终设计成一个独一无二的地标性娱乐景点，反映出法拉利的品牌特征和阿布扎比的雄心壮志。

阿布扎比法拉利主题公园于2010年12月开业，是全球最大的室内主题公园。Benoy设计的建筑具有法拉利高度可识别性的蜿蜒形式，其灵感来源于法拉利跑车底盘的双曲线轮廓。

建筑表现出一种"紧抱大地"的结构：像一个红色的沙丘。建筑的立体特征来源于法拉利跑车车身的双曲线，该双曲线结构同样应用在长700米、高45米的结构立面上。

金属屋顶有隔热功能，主立面使用高效玻璃来减少热负荷和强光照射。

在建筑的中央，屋顶像一个闪闪发光的水晶漏斗，为60米高的G-Force塔创造了完美的空间。

阿布扎比法拉利主题公园有20个娱乐项目，最新的模拟器可以让访客体验到法拉利赛车的激情及最高200千米/小时的速度。阿布扎比法拉利主题公园真是令人惊叹的体验中心。

UNIVERSITAT ROVIRA I VIRGILI

罗维拉·依维吉利大学

Architects: Ravetllat Ribas Arquitectes
Client: Universitat Rovira i Virgili (URV)
Location: Tortosa, Spain
Area: 983,580 m²
Photographer: Pedro Pegenaute

建筑机构：Ravetllat Ribas Arquitectes
客户：罗维拉·依维吉利大学
项目地址：西班牙托尔托萨
面积：983 580 平方米
摄影：Pedro Pegenaute

Halfway between the park and the city, the new building brings together several colleges and aims to make perfect transition between the nature and the urban. On the one hand, it can be discovered between the trees as a small fragmented piece allowing an easy integration with the surrounding park, but on the other hand, it is able to offer a more compact urban facade that binds with the urban topography reaching the usual height of the city.

The new campus wants to be the park's gate, something that gives meaning to the beginning or the end. Visible from the castle , the bridge is recognizable not by its size but its ability to respond to the different areas where it is located. From the top of the city the roof becomes a characteristic element. A roof visible in small dimensions aims, as another facade, to interact with the surrounding area.

The building's floor is fragmented to obtain the required perimeter for the program disposition and at the same time improve its circulation. Unlike a barrier building would represent, this distribution improves the diagonal itineraries and sets relations from the park towards the main street and from the city towards the new fairgrounds. The floor organization avoids conventional orthogonality after searching the use of two main directions derived from the context and the relationship with neighbor buildings.

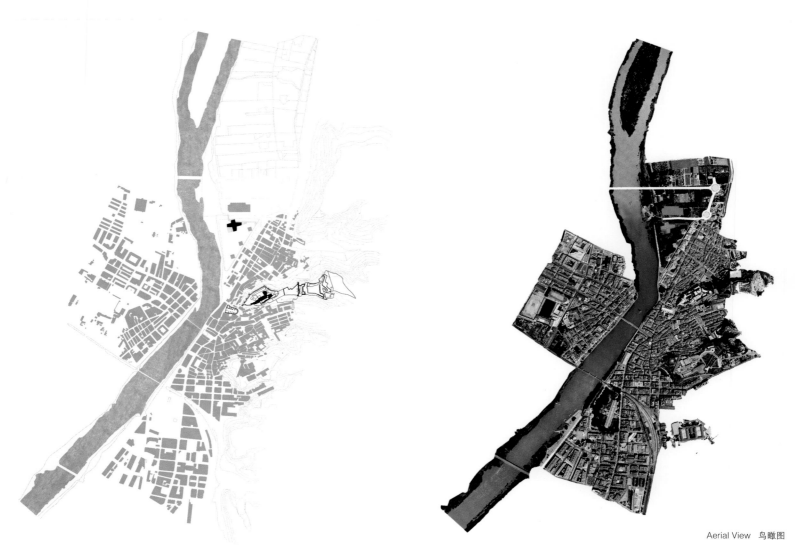

Aerial View 鸟瞰图

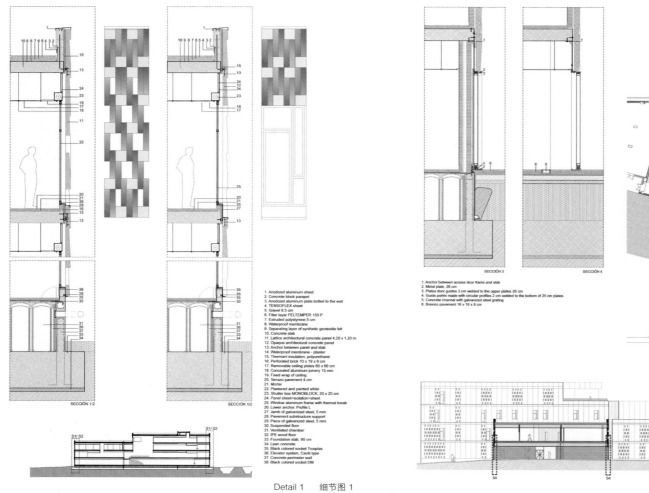

1. Anodized aluminum sheet
2. Concrete block parapet
3. Anodized aluminum plate bolted to the wall
4. TENSOFLEX sheet
5. Gravel 8.5 cm
6. Filter layer FELTEMPER 150 P
7. Extruded polystyrene 5 cm
8. Waterproof membrane
9. Separating layer of synthetic geotextile felt
10. Concrete slab
11. Lattice architectural concrete panel 4,20 x 1,20 m
12. Opaque architectural concrete panel
13. Anchor between panel and slab
14. Waterproof membrane - plaster
15. Thermani insulation, polyurethame
16. Perforated brick 10 x 19 x 9 cm
17. Removable ceiling plates 60 x 60 cm
18. Concealed aluminum joinery 15 mm
19. Fixed wrap of ceiling
20. Terrazo pavement 4 cm
21. Mortar
22. Plastered and painted white
23. Shutter box MONOBLOCK, 20 x 20 cm
24. Panel sheet+isolation+sheet
25. Window aluminum frame with thermal break
26. Lower anchor. Profile L
27. Jamb of galvanized steel, 5 mm
28. Pavement substructure support
29. Piece of galvanized steel, 5 mm
30. Suspended floor
31. Ventilated chamber
32. IPE wood floor
33. Foundation slab, 90 cm
34. Lean concrete
35. Black colored socket Trusplas
36. Elevator system, Cavtti type
37. Concrete perimeter wall
38. Black colored socket DM

1. Anchor between access door frame and slab
2. Metal plate, 26 cm
3. Plates door guides 3 cm welded to the upper plates 26 cm
4. Guide points made with circular profiles 2 cm welded to the bottom of 25 cm plates
5. Concrete channel with galvanized steel grating
6. Breinco pavement 16 x 16 x 8 cm

Detail 1　细节图 1

Detail 2　细节图 2

Its organization avoids the traditional scheme with long corridors serving rows of classrooms where there is usually a significant imbalance between the dimension of these corridors and their use. This type of organization can configure a good disposition of the program requested, grouping them by different subject areas, reducing the internal circulation, avoiding corridors and generating different places or small spaces for relation. These server spaces contain lockers, small offices, toilets, installations, etc., enhance the sound isolation and improve social activities. The double-axis and the variable geometry of the floor generate small squares and meeting places. The ceiling height and the variability of natural light through the proposed skylights contribute to the perception of these spaces as interior streets.

In section, the program which is more open and public stands on the ground floor, incorporating a hall, a space for exhibitions, a library connected with computer spaces, a study hall and a bar. The floors lose surface gradually, in coherence with the program but also trying to find an appropriate level of relationship with the park and generating outdoor terraces.

Outwards, we thought a modular system that ensures the continuity of the facade while fits different treatments for each orientation and specific needs of users in places where solar control is especially needed. This is the aspect that we want to highlight, in the way that we search for the continuity of the texture combining with the specificity of each orientation and programmatic requirements. The panels with variant holes are placed in front of the windows in most of the classrooms and offices. In big classrooms and other pieces with important dimensions one of the last panels is removed to allow a direct relationship with the outside. The constructive solution of the facade is determined as a ventilated facade with concrete panel in large format (4.20 m in height, 1.20 m in width).

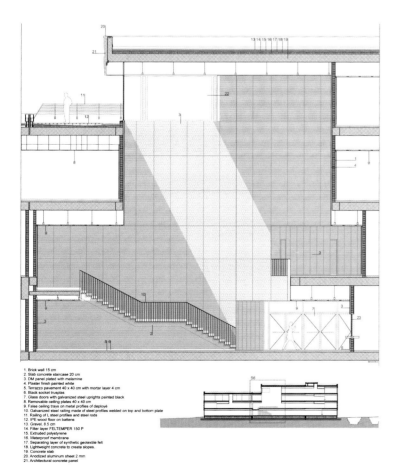

1. Brick wall 15 cm
2. Slab concrete staircase 20 cm
3. DM panel plated with melamine
4. Plaster finish painted white
5. Terrazzo pavement 40 x 40 with mortar layer 4 cm
6. Black socket trusplas
7. Glass doors with galvanized steel uprights painted black
8. Removable ceiling plates 40 x 40 cm
9. False ceiling trays on metal profiles of deployé
10. Galvanized steel railing made of steel profiles welded on top and bottom plate
11. Railing of L steel profiles and steel rods
12. IPE wood floor on battens
13. Gravel, 8.5 cm
14. Filter layer FELTEMPER 150 P
15. Extruded polystyrene
16. Waterproof membrane
17. Separating layer of synthetic geotextile felt
18. Lightweight concrete to create slopes
19. Concrete slab
20. Anodized aluminum sheet 2 mm
21. Architectural concrete panel

Detail 3　细节图 3

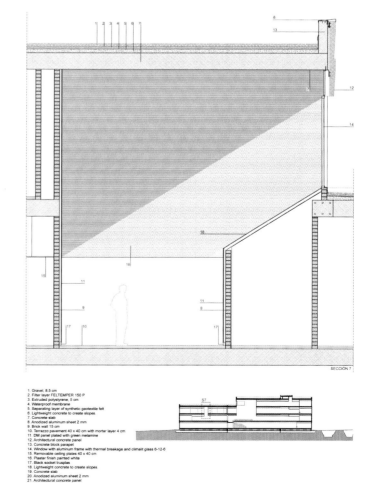

1. Gravel, 8.5 cm
2. Filter layer FELTEMPER 150 P
3. Extruded polystyrene, 5 cm
4. Waterproof membrane
5. Separating layer of synthetic geotextile felt
6. Lightweight concrete to create slopes
7. Concrete slab
8. Anodized aluminum sheet 2 mm
9. Brick wall 15 cm
10. Terrazzo pavement 40 x 40 cm with mortar layer 4 cm
11. DM panel plated with green melamine
12. Architectural concrete panel
13. Concrete block parapet
14. Window with aluminum frame with thermal breakage and climalit glass 6-12-6
15. Removable ceiling plates 40 x 40 cm
16. Plaster finish painted white
17. Black socket trusplas
18. Lightweight concrete to create slopes
19. Concrete slab
20. Anodized aluminum sheet 2 mm
21. Architectural concrete panel

Detail 4　细节图 4

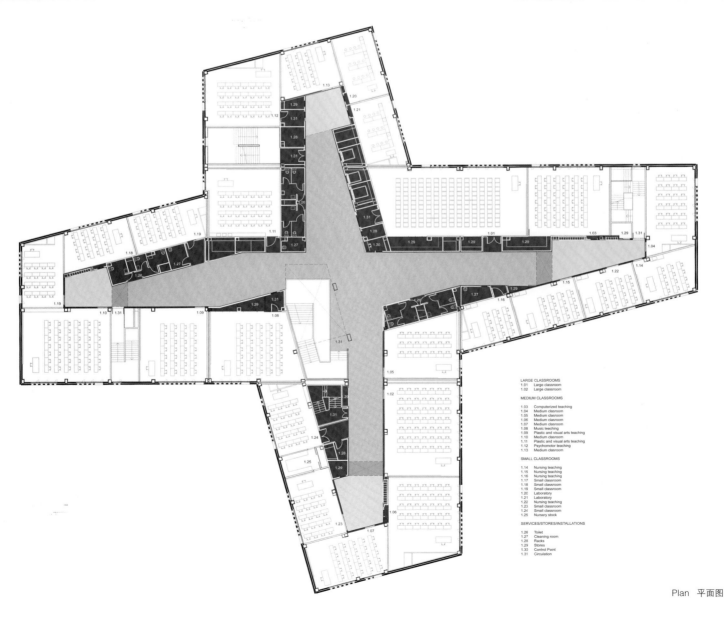

LARGE CLASSROOMS
1.01   Large classroom
1.02   Large classroom

MEDIUM CLASSROOMS

1.03   Computerized teaching
1.04   Medium classroom
1.05   Medium clasroom
1.06   Medium classroom
1.07   Medium classroom
1.08   Music teaching
1.09   Plastic and visual arts teaching
1.10   Medium classroom
1.11   Plastic and visual arts teaching
1.12   Psychomotor teaching
1.13   Medium classroom

SMALL CLASSROOMS

1.14   Nursing teaching
1.15   Nursing teaching
1.16   Nursing teaching
1.17   Small classroom
1.18   Small classroom
1.19   Small classroom
1.20   Laboratory
1.21   Laboratory
1.22   Nursing teaching
1.23   Small classroom
1.24   Small classroom
1.25   Nursery stock

SERVICES/STORES/INSTALLATIONS

1.26   Toilet
1.27   Cleaning room
1.28   Racks
1.29   Stores
1.30   Control Point
1.31   Circulation

Plan　平面图

这个新建筑位于城市和公园之间，旨在把周围几所大学连接起来，并在自然和城市之间创造完美的过渡。一方面，它像一个小型的建筑片段掩映在树木之中，与周围的公园融为一体。但另一方面，它紧凑简洁的城市立面与城市地貌和高度相协调。

新的大学校园旨在成为公园的大门，给予其开始或终结的含义。在城堡上可以看到桥梁，桥梁与周围景观相呼应。从城市的高处俯瞰建筑，屋顶是它的特色所在，作为建筑的另一个立面，屋顶与周围环境相呼应。

建筑的楼层被分隔以满足项目布局不同周长的要求，同时也促进了人流流通。这种布局促进了对角行程，在城市和主街道、城市和新露天广场间形成联系，而不是形成障碍。在研究和周围环境紧密相关的两个主要方向以及与周围建筑的关系之后，楼层没有采用传统的正交布局。

项目方案避免了传统方案中长走廊及成排教室的模式，这些走廊在使用时通常有不平衡之处。这种类型的布局可以很好地满足项目的需求。将它们按不同的领域分组，减少内部循环，避免走廊过多，增加不同的小空间用于社交。这些小的服务空间包括寄物柜、小办公室、厕所、设施间等，加强了隔音效果，促进了社交活动的开展。楼层的双轴系统和可变的几何图形产生了小型广场和聚会场所。天花板的高度和通过天窗透过的可变自然光线加强了这些空间的可识性，它们就像一个个的内部街道。

首层的空间更加开放，为大众服务，这里有一个大厅、展览空间、和计算机房相连接的图书馆、学习大厅以及一个酒吧。楼层的表面逐渐消失，与建筑体相一致。此外，它尝试与公园产生恰当的关系并形成一个室外的平台。

在外部，设计师设计了一个模块化系统，可以确保立面的连贯性，使每个方向都得到不同处理，并满足需要日照控制系统地方的使用者的需求。这是设计的亮点。设计师结合每个方向的特征和项目的规划需求，探索具有连贯性的结构。在大多数教室和办公室的窗户前边均设置带有孔洞的面板。在大教室和其他的重要空间则不用面板，而是与户外直接相连。立面采用了建设性的方案，使用大型混凝土板（4.2米高，1.2米宽）形成通风立面。

Section 剖面图

THE MEDIAPRO BUILDING AT THE AUDIOVISUAL CAMPUS

MEDIAPRO 视听校园大楼

Architects: CARLOS FERRATER – XAVIER MARTI – PATRICK GENARD
Client: MEDIACOMPLEX S.A.
Collaboration: DARIELA HENTSCHEL
Structural Design: JUAN CALVO (PONDIO)
Location: NEW AV. DIAGONAL IN 22@ DISTRICT BARCELONA
Area: 52,456m²
Photographer: Alejo Bagué

设计机构：CARLOS FERRATER–
XAVIER MARTI–PATRICK GENARD
客户：MEDIACOMPLEX S.A.
合作机构：DARIELA HENTSCHEL
结构设计：JUAN CALVO (PONDIO)
项目地址：NEW AV. DIAGONAL
　IN 22@ DISTRICT BARCELONA
面积：52 456 平方米
摄影：Alejo Bagué

Urban system

First piece is in a sequence with a perspective view of the Torre Agbar. In its "first piece" role it responds to the geometry of the diagonal by "turning its head".

This turn leads to a displacement with the podium of the adjacent tower and the jut over the piazza, and constructs the front and side elevations. The fusion of the two geometries opens up the sightline of the TMB building.

The low, transparent body frees up the Calle Bolivia and presents a "lobby-style" public space as a continuation of the piazza and by means of an also-continuous paving.

Small modifications are made in the podium programme in order to adjust the imbrication of the two volumes; in this way the facade plane penetrates below the tower, thus prolonging the urban space of the Calle Bolivia. The crown perpendicular to the diagonal conceals the installations and defines the rear arris of the intervention.

城市系统

在整个系列的第一部分有一个 Torre Agbar 的透视图，在它的第一部分里通过"调转顶部"来反映其对角线的几何原理。

这种调转导致了相邻建筑墩座之间的位移以及广场的突起，同时构成了正面和侧面的高度。两种几何形状的融合打开了 TMB 建筑的视线。

在低处透明的主体可以看到玻利维亚大道，它那"大堂式风格"的公共区域表现成为广场的延续，并且使用了同样能表现延续性的铺筑材料。

为了配合两种体块的鳞形结构，小的改动被安排在墩柱制作程序上。这样一来，表皮面穿过建筑底部，延长了玻利维亚大道的城市空间。顶点垂直于对角线的结构不仅隐藏了装置，并且决定了后棱的介入。

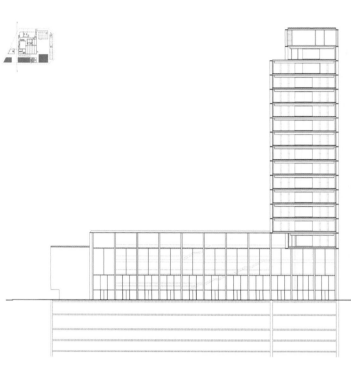

Elevation 1 立面图 1

Elevation 2 立面图 2

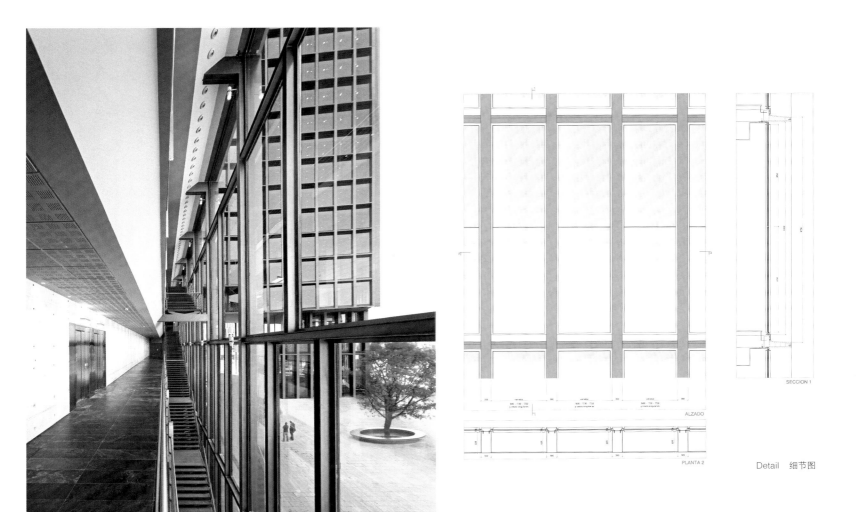

ALZADO

PLANTA 2

SECCION 1

Detail 细节图

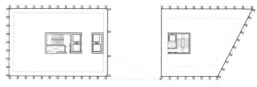

Planta Baja

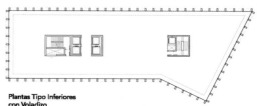

Plantas Tipo Inferiores
con Voladizo

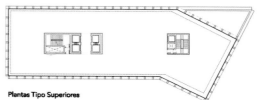

Plantas Tipo Superiores

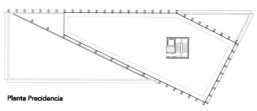

Planta Precidencia

Plan 平面图

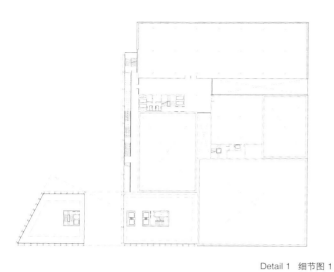

Detail 1 细节图 1

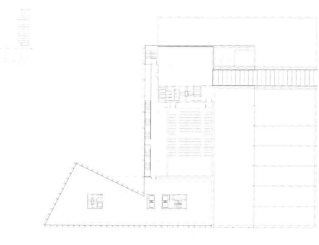

Detail 2 细节图 2

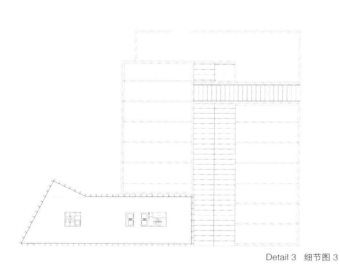

Detail 3 细节图 3

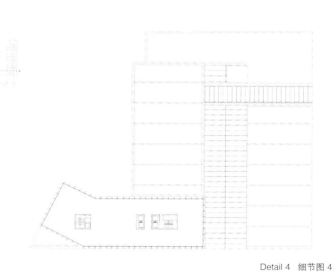

Detail 4 细节图 4

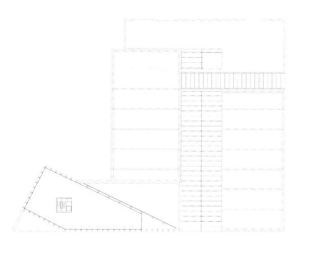

Detail 5 细节图 5

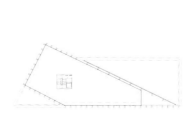

Detail 6 细节图 6

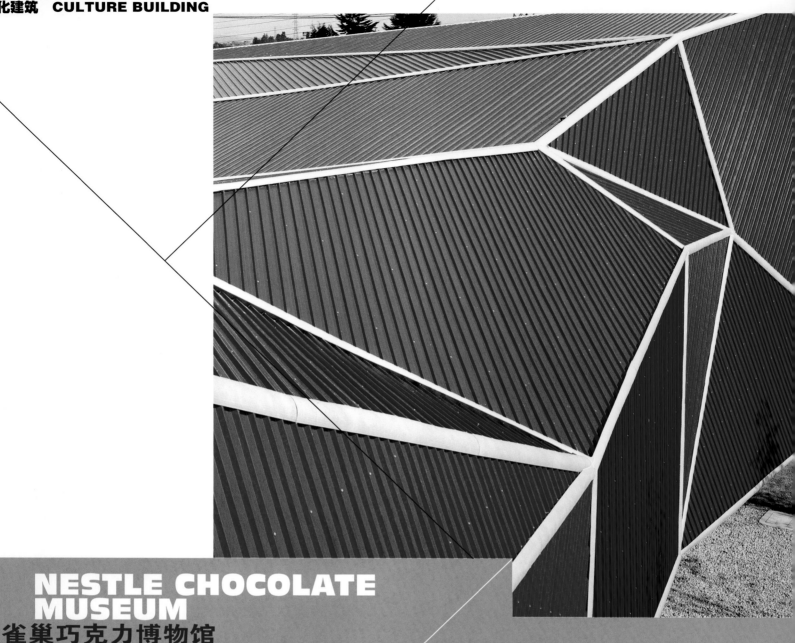

NESTLE CHOCOLATE MUSEUM
雀巢巧克力博物馆

Architects: Rojkind Arquitectos
Design Leader: Michel Rojkind
Location: Toluca, Mexico
Area: 634 m²
Photographer: Guido Torres, Paúl Rivera

设计机构：Rojkind Arquitectos
项目总监：Michel Rojkind
项目地址：墨西哥托卢卡
面积：634 平方米
摄影：Guido Torres, Paúl Rivera

Project Team: Agustin Pereyra, Mauricio
Garcia-Noriega, Moritz Melchert, Juan Carlos
Vidals, Paulina Goycoolea,
Daniel Dusoswa, Matthew Lohden
Lighting Designer: Noriega Arquitectonics
Iluminators [Ricardo Noriega], Fernando Gonzáles
Landscape Designer: Ambiente Arquitectos & Asociados
[Fritz Sigg, Juan Guerra], Erick Flores

项目团队：Agustin Pereyra, Mauricio Garcia-Noriega,
Moritz Melchert, Juan Carlos Vidals,
Paulina Goycoolea, Daniel Dusoswa, Matthew Lohden
灯光设计：Noriega Arquitectonics
Iluminators [Ricardo Noriega], Fernando Gonzáles
景观设计：Ambiente Arquitectos & Asociados [Fritz Sigg, Juan Guerra],
ErickFlores

Architecture as an experience. Sensory architecture, experienced through the architectural tour, through the surprises, the turns and the bends. Architecture as a challenge. The forms and spaces contained, as well as the times are pushed to the limit. Complexity and record time: three months to project and build.
Located over the side lane of the highway in the entrance to Toluca, in the edge of a 300 meter-long insubstantial industrial installation that used to pass unnoticed, the new object appears with the spectacular nature of a window display. Half way between Mathias Goeritz's The Snake and Munch's Scream, this zigzagging origami rises from the garden level and becomes the entrance to a magical world, to the tour of the chocolate factory that rivals Tim Burton's imagination. The 600 m² new construction standing over the garden, houses a reception area, a theater that prepares young visitors for the trip to the world of chocolate; the entry to the existing tunnel that circles around the production areas is in the inside of the factory and the chocolate and gadget store is at the end of the tour.
The triangles of the unfolding kaleidoscope are made out in different shades of white to accentuate the different planes. The lobby opens up over an insipid view of high voltage cables, and billboards and highway give way to visitors between the information desk and the chocolate-bar shaped sofas. The theater in this little EPCOT, encloses the visitors for a few minutes to introduce them virtually to the liquid world of candy. From there, the tour begins through the corridors, tunnels and observation decks over the halls of the factory. Before leaving, a store invites us to perpetuate the moment with objects to take home and thrones that transform us into royalty for an instant.
This urban-scale toy invites us on an emotional tour and gives free rein to the exuberant creativity. The new alebrije-red on the outside and white on the inside made from urgent origami, bursts out like a unique icon in the Tolucan periphery.

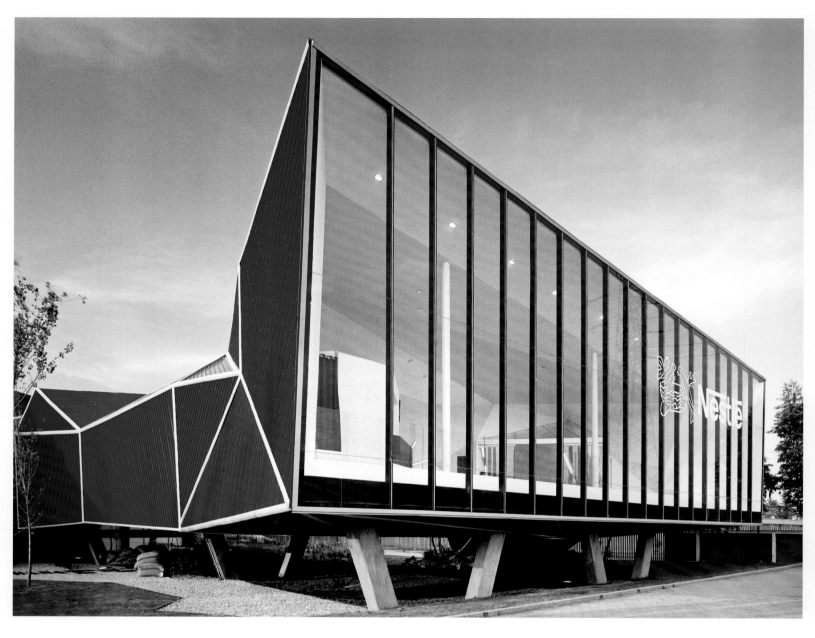

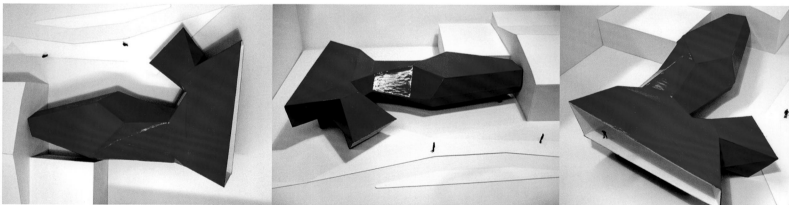

Modelling 模型图

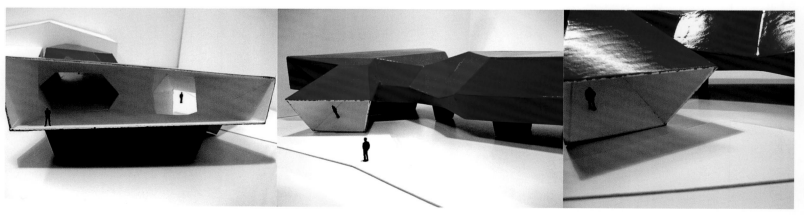

在这里，建筑是一种体验，你可以体验到惊奇的建筑之旅。在这里，建筑是一种挑战，建筑形式、空间、时间都突破了限制。项目设计及建造完成仅用时 3 个月。

此建筑位于高速公路在托卢卡入口的外侧车道上，在一个不易被人注意的 300 米长的工业基地边缘，新建筑的橱窗设计体现了其独特之处。在 Mathias Goeritz 的 The Snake 和 Munch's Scream 之间，一个弯曲的折纸结构耸立在花园中，成为通往神奇世界的一个入口，开启了一个与蒂姆·伯顿想象中不同的巧克力工厂之旅。600 平方米的新建筑耸立在花园上，这里有接待区和一个剧场，为准备进入巧克力世界的年轻游客提供服务。工厂内部环绕生产区的通道入口位于工厂内，巧克力和小商品商店则在通道的尽头。

万花筒上的三角形由不同色调的白色衬托以突出不同的平面。大堂在高压线上方对外开放，广告牌和公路为在服务台和长巧克力状的沙发之间的人们让路。剧院则可以为游客介绍数分钟的液体糖果世界的知识。从那里，游客可以穿过走廊、隧道和工厂大厅上方的观察平台。在离开之前，游客可以去商店购买些礼物带回家，同时让人们体验一下尊贵的感觉。

这个城市规模的玩具邀请我们体验一次激情之旅，感受旺盛的创造力。这个外红内白的折纸建筑已成为托卢卡区的新地标。

(A) East View
(B) North View
(C) South View
(D) West View

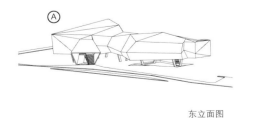

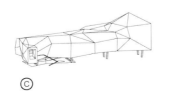

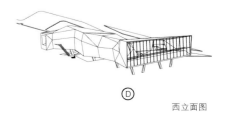

东立面图　　　　　　　　北立面图　　　　　　　　南立面图　　　　　　　　西立面图

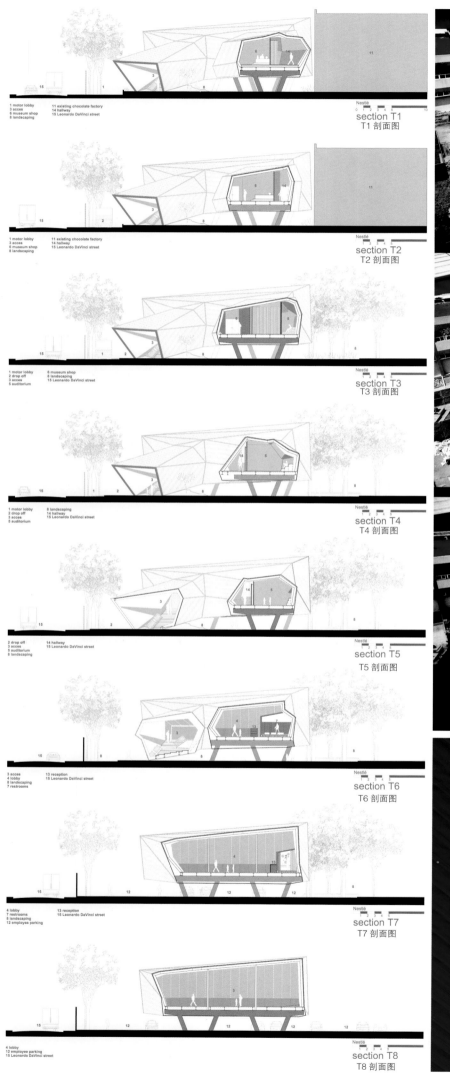

1 motor lobby
3 acces
6 museum shop
8 landscaping
11 existing chocolate factory
14 hallway
15 Leonardo DaVinci street

Nestlé

section T1
T1 剖面图

1 motor lobby
3 acces
6 museum shop
8 landscaping
11 existing chocolate factory
14 hallway
15 Leonardo DaVinci street

Nestlé

section T2
T2 剖面图

1 motor lobby
2 drop off
3 acces
5 auditorium
6 museum shop
8 landscaping
15 Leonardo DaVinci street

Nestlé

section T3
T3 剖面图

1 motor lobby
2 drop off
3 acces
5 auditorium
8 landscaping
14 hallway
15 Leonardo DaVinci street

Nestlé

section T4
T4 剖面图

2 drop off
3 acces
5 auditorium
8 landscaping
14 hallway
15 Leonardo DaVinci street

Nestlé

section T5
T5 剖面图

3 acces
4 lobby
8 landscaping
7 restrooms
13 reception
15 Leonardo DaVinci street

Nestlé

section T6
T6 剖面图

4 lobby
7 restrooms
8 landscaping
12 employee parking
13 reception
15 Leonardo DaVinci street

Nestlé

section T7
T7 剖面图

4 lobby
12 employee parking
15 Leonardo DaVinci street

Nestlé

section T8
T8 剖面图

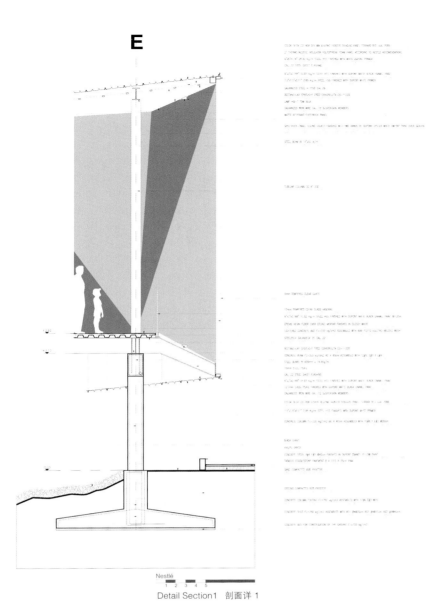

E

Nestlé
1 2 3 4 5

Detail Section1　剖面详 1

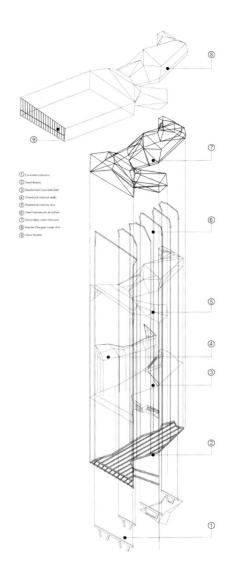

① Concrete columns
② Steel Beams
③ Reinforced Concrete Slab
④ Sheetrock interior walls
⑤ Sheetrock interior skin
⑥ Steel framework structure
⑦ Secondary steel structure
⑧ Hunter Douglas outer skin
⑨ Glass facade

Detail 1　细节图 1

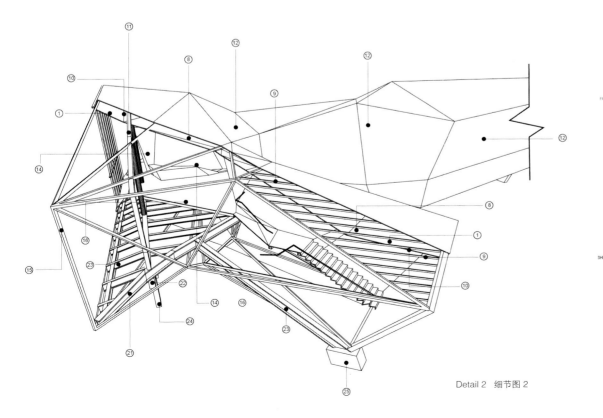

Detail 2　细节图 2

① SHEETROCK PANEL CEILING CALKED FINISHED WITH TWO HANDS OF DUPONT OYSTER WHITE OW79P PAINT OVER SEALANT
PLAFÓN A BASE DE PANEL DE TABLAROCA CALAFATEADO ACABADO A DOS MANOS DE PINTURA ACRÍLICA DUPONT BLANCO AMANECER OA68P SOBRE SELLADOR

② GALVANIZED IRON WIRE CAL. 12 SUSPENSION MEMBERS
TIRANTES DE ALAMBRE GALVANIZADO CAL. 12

③ GALVANIZED STEEL H STUD CAL 26
CANAL LISTÓN DE LÁMINA GALVANIZADA CAL. 26

④ DIURETAN PERIMETRAL SEAL
SELLO PERIMETRAL CON DIURE TAN

⑤ STEEL SHEET CAL 22 ROLLED ACCORDING TO DESIGN, NATURAL ALUMINUM COLOR cod. 7163
LÁMINA CAL. 22 ROLADO SEGÚN DISEÑO COLOR SATIN NATURAL cod. 7163

⑥ "J" EDGE COVER STEEL SHEET CAL. 22 ROLED ACCORDING TO DESIGN
"J" REBORDE A BASE DE LÁMINA CAL. 22 ROLADO SEGÚN DISEÑO

⑦ 6"x6"X1/4" 28.30 Kg/m STEEL HSS FINISHED WITH WHITE DUPONT PRIMER
HSS DE ACERO 6"x6"X1/4" 28.30 Kg/m ACABADO CON PRIMER DUPONT BLANCO

⑧ 2" THERMO/ACUSTIC INSULATOR POLYSTYRENE FOAM PANEL ACCORDING TO NESTLE RECOMENDATIONS
PANEL DE ESPUMA DE POLIESTIRENO DE 2" COMO AISLANTE TERMO/ACÚSTICO SEGÚN RECOMENDACIONE S DE NESTLE

⑨ 1.5"x1.5"x0.11" 2.95 Kg/m STEEL HSS FINISHED WITH DUPONT WHITE PRIMER
HS S D E ACERO 1.5"x1.5"x0.11" 2.95 Kg/m ACABADO CON PRIMER DUPONT BLANCO

⑩ WATER RESISTANT SHEETROCK PANEL
PANEL DE TABLAROCA RESISTENT E A LA HUMEDAD

⑪ STEEL BEAM IR 16"x50 lb/ft
VIGA DE ACERO IR 16"x50 lb/ft

⑫ COLOR DECK CD-408 0.5mm ALUZINC HUNTER DOUGLAS PANEL FERRARI RED cod. 7088
PANEL COLOR DECK CD-408 DE ALUZINC DE 0.5mm HUNTER DOUGLAS COLOR ROJ O FERRARI cod. 7088

⑬ GALVANIZED STEEL C STUD CAL 26
POST E "C" DE LÁMINA GALVANIZADA CAL. 26

⑳ 1" STEEL PLATE
PLAC A DE ACERO DE 1"

⑮ 6"x6"X1/4" 28.30 Kg/m STEEL HSS FINISHED WITH WHITE DUPONT PRIMER
HSS DE ACERO 6"x6"X1/4" 28.30 Kg/m ACABADO CON PRIMER DUPONT BLANCO

⑯ EPOXIC RESIN FLOOR OVER EPOXIC MORTAR FINISHED IN GLOSSY WHITE
PISO DE RESINA EPÓXICA COLOCADO SOBRE MORTERO EPÓXICO ACABADO EN BLANCO BRILLANTE

⑰ LIGHTENED CONCRETE BED fc=200 kg/cm2 ASSEMBLED WITH 6X6-10/10 ELECTRO-WELDED MESH
FIRME DE CONCRET O ALIGERADO f'c=200 kg/cm2 ARMADO CON MALLA ELECTRO-SOLDADA 6X6-10/10

⑱ STEELDECK GALVADECK 25 CAL. 22
LOSACERO GALVADECK 25 CAL. 22

⑲ STEEL BEAM IR 457mm x 144.3Kg/m
VIGA DE ACERO IR 457mm x 144.3Kg/m

⑭ SHEETROCK PANEL WALL CALKED FINISHED WITH TWO HANDS OF DUPONT OYSTER WHITE OW79P PAINT OVER SEALANT
MURO A BASE DE PANEL DE TABLAROCA CALAFATEADO ACABADO A DOS MANOS DE PINTURA ACRÍLICA DUPONT BLANCO OSTIÓN OW79P SOBRE SELLADOR

㉑ STEEL BEAM IR 406mm x 74.4Kg/m
VIGA DE ACERO IR 406mm x 74.4Kg/m

㉒ CONCRETE BEAM fc=350 kg/cm2 40 X 80cm ASSEMBLED WITH 12#8, 2#4 Y E#4
TRABE DE CONCRET O ARMADO f'c=350 kg/cm2 DE 4 0 X 80cm CON 12#8, 2#4 Y E #4

㉓ 6"x6"X1/4" 28.30 Kg/m STEEL HSS FINISHED WITH WHITE DUPONT PRIMER
HSS DE ACERO 6"x6"X1/4" 28.30 Kg/m ACABADO CON PRIMER DUPONT BLANCO

㉔ CONCRETE COLUMN fc=350 kg/cm2 60 X 40cm ASSEMBLED WITH 10#6 Y E#3 @20cm
COLUMNA DE CONCRET O ARMADO f'c=350 kg/cm2 DE 6 0 X 40cm CON 10#6 Y E#3 @20cm

㉕ CONCRETE COLUMN FOOTING fc=250 kg/cm2 ASSEMBLED WITH 12#6 E#3 @20
DADO DE CONCRET O ARMADO f'c=250 kg/cm2 CON 12#6 E#3 @ 20

㉖ CONCRETE SHOE fc=250 kg/cm2 ASSEMBLED WITH AS1-#4@20cm AS2-#4@25cm AS3-#4@40cm
ZAPATA DE CONCRET O ARMADO f'c=250 kg/cm2 AS1-#4@20cm AS2-#4@25cm AS3-#4@40cm

㉗ CONCRETE BED FOR CONSOLIDATION OF THE GROUND fc=100 kg/cm2
FIRME DE CONCRET O ARMAD O f'c=100 kg/cm2

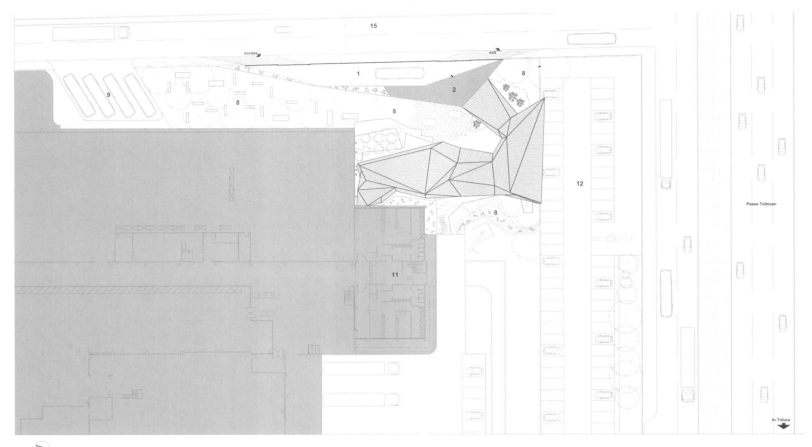

Site Plan 总平面图

1 motor lobby
2 drop off
8 landscaping
9 school bus parking

11 existing chocolate factory
12 employee parking
15 Leonardo DaVinci street

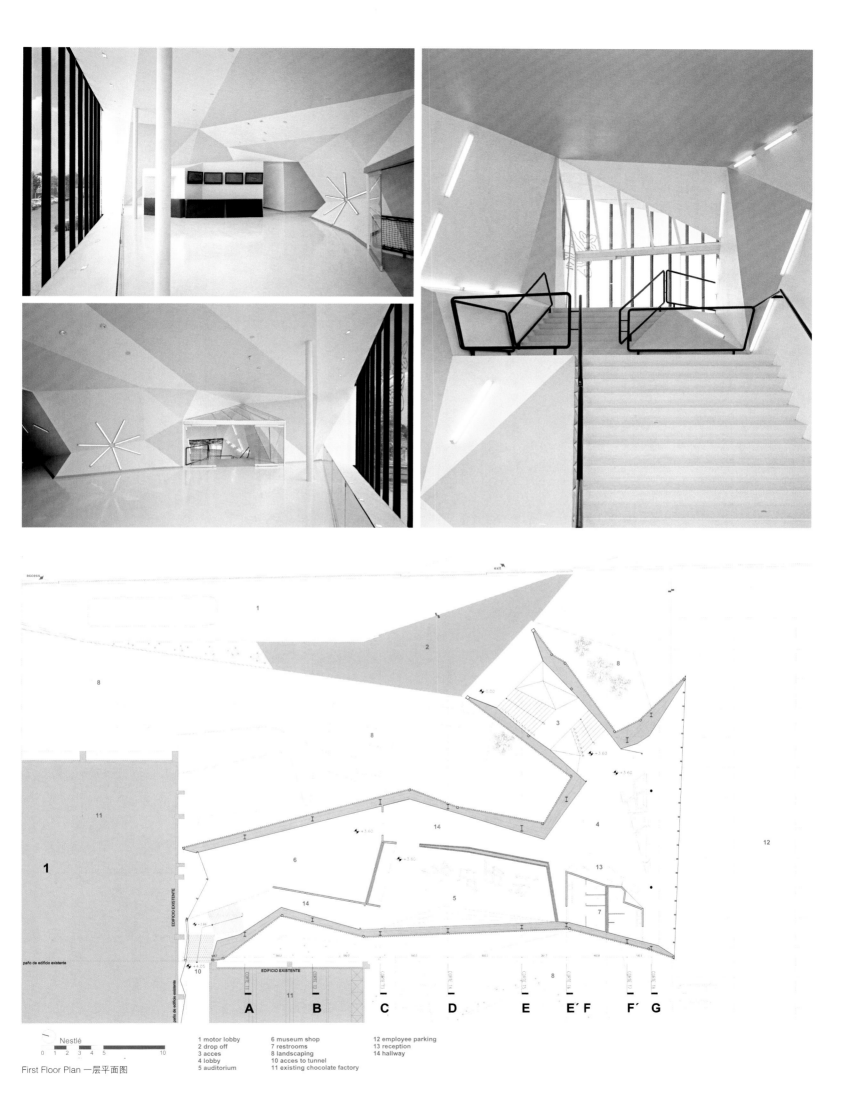

First Floor Plan 一层平面图

Nestlé
0 1 2 3 4 5   10

1 motor lobby
2 drop off
3 acces
4 lobby
5 auditorium
6 museum shop
7 restrooms
8 landscaping
10 acces to tunnel
11 existing chocolate factory
12 employee parking
13 reception
14 hallway

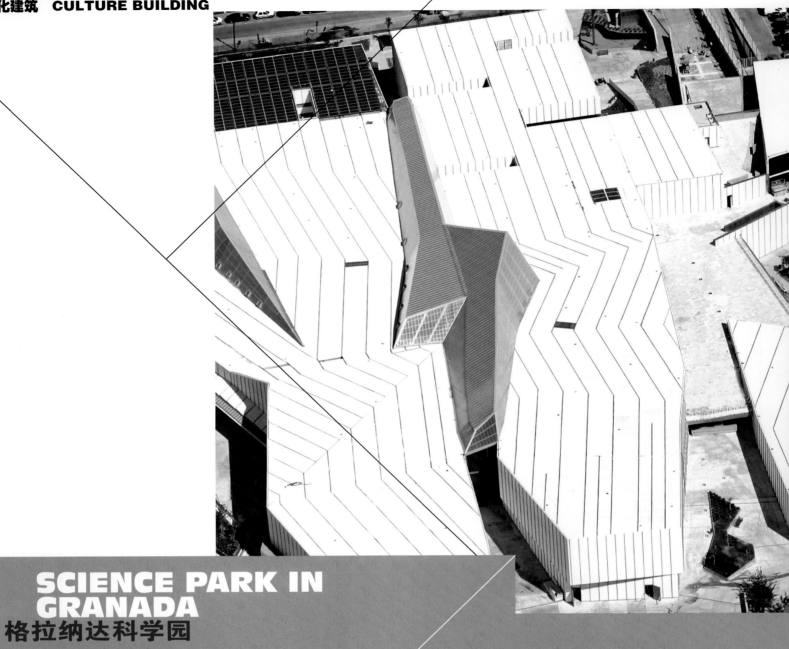

SCIENCE PARK IN GRANADA
格拉纳达科学园

Architects: Carlos Ferrater – Eduardo Jimenez, Yolanda Brasa
Client: Consorcio Parque Delas Ciencias
Structural Design: Juan Calvo (Pondio)
Constructor: DRAGADOS
Location: Granada, Spain
Area: 48,377 m²
Photographer: Alejo Bague

设计机构：Carlos Ferrater – Eduardo Jimenez, Yolanda Brasa
客户：Consorcio Parque Delas Ciencias
结构设计：Juan Calvo (Pondio)
建筑公司：DRAGADOS
项目地址：西班牙格拉纳达
面积：48 377 平方米
摄影：Alejo Bague

The project for extending the Science Park in Granada, next to the River Genil, posits the construction of a single roof with slight inflections that is similar to an open hand, beneath which are housed the different bits of the program interwoven in a spatial continuum.

The empty space connecting the big boxes or containers of the different programs—the Macroscope, Biodome, Technoforum, Health Sciences, Al-Andalus, auditoria and spaces for temporary and permanent exhibitions—organizes the communications and logistics and relates to the Park's activities as a whole.

The proposed spatial structure allows for great flexibility in terms of uses and of situations interweaving different routes and themes.

In general, most of the projects that develop topographies replace the continuous quality of the roof with a succession of planes or porticos of various shapes. In this way, the spatiality and constructional autonomy of the roof as a continuous element is converted into a sequential relationship of intercommunicating spaces.

In Granada the roof constitutes a continuous angled plane that floats above the sloping plane of the ground, the two planes enclosing the huge exhibition spaces between them, with the light that penetrates between the roof's angles emphasizing the places of communication and relation.

In its abstraction, the great roof displays a silhouette that recalls the skyline of the mountains of Granada. The resolution of the great topographical roof adapts to volumetric needs, which generate within them closed spaces of great size and height that accommodate the different programs. The roof hovers above the terrain, constructing a new topography that, by being angled, organizes in its angles the skylights that endow the spaces of connection and circulation with natural light.

The grid of the roof is resolved with a double-layered, three-dimensional structure that includes technical systems and installation networks, thus solving drainage and rain run-off. The skylights punctuate the roof as a continuation of the main structure.

Elevation 立面图

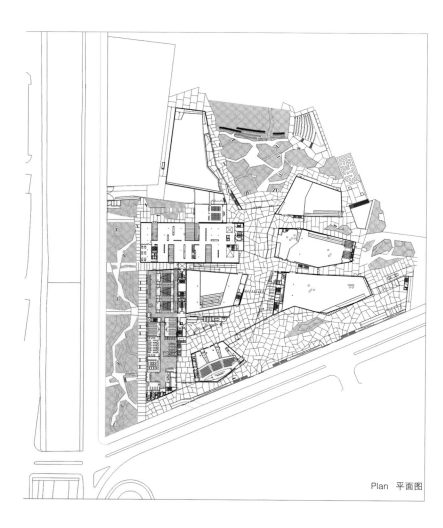

Plan 平面图

格拉纳达科学园扩展项目紧邻赫尼尔河，搭建出一个轻微折叠的单一屋檐，类似于一个张开的手掌，下面是一个涵盖不同功能、相互关联的连续空间。

架空层连接不同的功能区，将交通和后勤贯穿起来，并且将园区的活动连成一体。

规划的空间结构考虑到灵活性，以应对不同路径和主题下的使用以及实际情况。

一般来说，大部分工程因地制宜，用一连串的截面或各种形状的门廊取代连绵的屋顶。这样一来，屋顶的空间性和结构的主导性作为一个不变的成分，运用到关联空间的顺序关系中。

在格拉纳达，屋顶构成一个连续的成角平面，立浮于地面斜坡之上，两个平面将大型展览空间包围其中，光线从屋角穿透而入，突显场馆的交通与联系。

抽象来讲，屋顶显现的剪影让人想起格拉纳达山脉的轮廓。屋顶设计运用地形学原理以满足容量的需求，从而在内部形成大规模的封闭空间以及能容纳各种活动的场馆高度。屋顶依势而起，通过角度营造新的地势条件，倾斜而开的天窗确保了自然光在空间内的畅通。

屋顶具有双层三维结构网格，包含技术系统和网络配置，以解决雨水疏散的问题。点缀于屋顶之上的天窗作为主体结构的延续而存在。

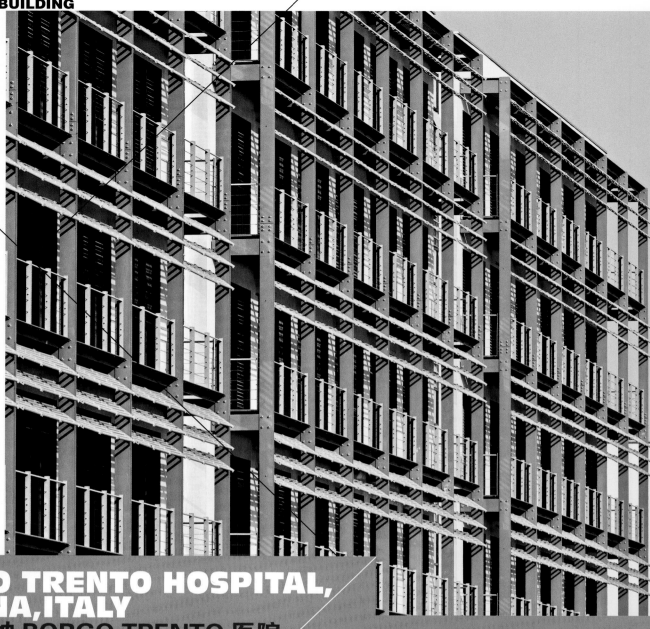

BORGO TRENTO HOSPITAL, VERONA, ITALY
意大利维罗纳 BORGO TRENTO 医院

Architects: gmp · von Gerkan, Marg and Partners with Studio Altieri
Associated Architects: Studio cfk, Clemens Kusch and Martin Weigert
Client: Azienda ospedaliera di Verona
Project Leaders: Robert Friedrichs, Arne Starke
Location: Verona, Italy
Building Area: 96,300 m²
Photographer: Marcus Bredt

设计机构：gmp 建筑师事务所
合作机构：Studio cfk, Clemens Kusch and Martin Weigert
客户：Azienda ospedaliera di Verona
项目负责人：Robert Friedrichs, Arne Starke
项目地址：意大利维罗纳
建筑面积：96 300 平方米
摄影：Marcus Bredt

Borgo Trento Hospital is of general importance for northern Italy. It is a pilot project by the Italian Ministry of Health, involving a thorough update and modernization of a large working inner-city hospital while keeping all hospital functions running throughout the construction period.

Located by the River Adige in north Verona close to the city centre, the hospital is getting a new multi-functional, compact complex of buildings accommodating various medical and general functions. These include surgery, intensive care, wards, operation theatres, Accidents & Emergencies (A&E), clinics and outpatients' department, and radiology, plus public areas (shops, catering etc.).

The new block now completed in the first phase has nine storeys overall, and is in the centre of the century-old hospital site. The existing buildings in this area were demolished, and their functions—previously distributed over a motley of individual conversions and additional buildings—have been integrated in the new buildings. This enables the operation of the hospital to be restructured for future needs, and considerably improves operations functionally and drives down costs (shorter transits, better care, maintenance, etc.).

The second phase will see vital functional areas such as laboratories and analysis installed in the Piastra, a two-storey building below ground level, lit by natural daylight by means of sunken garden courtyards. Putting these facilities below ground, in combination with the compactness of the high-rise block, frees up space for a new landscaped, park-like area called the Great Garden.

As a key design feature, the Great Garden in the centre of the hospital site establishes a buffer zone between old and new. More importantly, it provides a large, high-quality park area enhanced by architectural features (pergola, fountains, steps, ramps, bridges etc.), where visitors and patients can relax in the open air.

The new buildings at Borgo Trento consist basically of four components:
• the Polo, a square main building with an interior courtyard. Inter alia, it includes 33 operating theatres (the biggest such facility in Italy), surgery, intensive care and wards on the top three floors and an atrium with shops, catering etc. on the ground floor;
• the Ambulatorium, a four-storey wing in front of the Polo containing the main entrance lobby and specialist medical areas of the polyclinic and outpatients' department (with their own operating theatres);
• the Pronto Soccorso, a two-storey group likewise in front of the Polo and facing the river. This handles A & E, and is accessed from the riverside road. It also has direct access to the operating area.
In addition to the above buildings, a new technical facility was constructed in a different part of the hospital site, the design scheme of which following that of the overall site;
• in a later second phase comes the Piastra, a below-ground two-storey building with a diagnostic and therapy centre (Level-1: radiology, transfusions, physiotherapy, laboratories, etc.; Level -2: technical), with an enclosing access route linking with existing buildings in the area.

Elevation 立面图

THE WORLD ARCHITECTURAL
FIRM SELECTION

Borgo Trento 医院在意大利北部具有重要意义。这是意大利卫生部的一个试点项目，包括对一个大型的内城医院进行彻底的更新和现代化改造，并且在项目实施期间要保持医院能正常运行。

医院坐落于维罗纳北部、阿迪杰河畔、紧邻城市中心，是一个新建的综合建筑群体，各医学学科部门以及日常功能单位均安置于内（包括外科、重症监护室、病房、手术室、急诊、门诊、放射科以及店铺、餐饮等公共设施区）。

一期工程中建于已有百年历史的医院基地上的一栋九层高的住院楼现已完工。原有的旧建筑已被拆除，先前分布于不同单体建筑的多种功能将集中安置于新建建筑内。这使得医院的经营管理模式将更加适应未来的发展趋势，并在功能架构以及经济性能上更具优势（更便捷的交通，更好的护理与维护）。

二期工程包括很多重要的功能区，例如实验室和分析室，它们被布置在一个名为 Piastra 的建筑中，这是一个两层的下沉建筑，其日常采光通过下沉的中央庭院实现。通过这种设计手法，在集约紧密的高层建筑群中，一片绿化园林景观——"新庭园"应运而生。

新庭园作为设计的中心理念，一方面构成了医院新旧建筑之间的过渡区域，另一方面通过运用建筑元素（蔓藤回廊、喷泉、阶梯、坡道、桥梁等）塑造出宏伟大气、舒适宜人的园林空间，可供病患以及来访者在户外休憩。

Borgo Trento 医院的新建筑群主要由四栋建筑组成。

· Polo 是一座拥有中庭的立方体主楼，包括 33 间手术室（意大利最大的手术中心），外科、重症监护室、病房位于建筑上部三层，首层设有店铺以及餐饮设施。

· "门诊部"为一栋四层高的建筑，坐落于 Polo 前，其内设有入口门厅以及门诊部（包括独立的手术室）。

· Pronto Soccorso 是一座面向阿迪杰河的两层建筑，同样为于 Polo 主楼之前，其内设有急诊处，可将通过河岸大街到达的病患直接送至手术区。

除上述功能区域外，医院基地的另一侧还新建有一个技术支持中心，其规划方案与整体方案相一致。

· 在未来的二期工程中将建成 Piastra，一座两层高的下沉建筑，其中设有诊疗中心（地下一层：放射科，注射科，理疗室，实验室等；地下二层：技术设备间）以及一个围绕建筑的封闭通道，建立与周边建筑的联系。

综合建筑 COMPLEX BUILDING

After the era that the distinction between the residential real estate and commercial real estate is strict,the single real estate development mode could no longer satisfy the residency requirements of people's lives,also could not adapt to the rapid development of the real estate.Creating a new real estate development mode has become a new issue that developers need to facing. Mixed–using building integrates many functions of the city, including commerce, office, residence, hotel, exhibition, catering, conference, entertainment and transportation, and establishes a dynamic relationship of interdependence in various parts,then forming a versatile, high–efficiency building community. As the urban mixed–using building has all functions of modern city, they are often called as "city within a city".

A successful mixed–using building need to complete with four points.First of all, it needs to be creative and correspond with the external shape of the era aesthetics. Secondly, it needs to make good mixed–using planning, scientific and reasonable arrangement of residence, hotel–style apartments, office buildings, shopping malls, hotels, theme parks and other projects. However, these depend on the local urban development process, family income, residents' purchasing power and consumption structure. Thirdly, it needs to reasonably arrange the stream of people, traffic organization, and combination of space, etc. Different industries do not interfere with each other, making complex really play a compact, comprehensive role. Finally, it needs to pay attention to the combination of architecture and ecology, to increase the green area and achieve humanistic ecology. So all kinds of people in the complex could live together with each other harmoniously.

Mixed–using buildings are not only the landmark buildings of city, but also have been the standard of international life style system of each big city's commercial center district.They are the symbol of high quality city life, they can improve the city's overall image quality and value.The complex of mixex–using building determines that it has strong social function which is the engine of regional economy and the main factors that could enhance urban economy and culture. Excellent commercial complex can enhance the flow of the city, improve the quality of life and consumption,and become an important tourist attraction in the city. All these can enhance the attraction of the city.

The development of the modern city is transformed from extensive to intensive direction. Urban mixed–using buildings are more and more emphasizing on the city's openness and integration, focusing on the construction of urban public space, which forms city's public spaces, such as the transportation hub, cultural square.Through the linkage with city, regional development has been integrated.

经历了将住宅地产和商业地产严格区分的时代后，单一的地产开发模式已经渐渐不能满足人们对生活居住的要求，也逐渐跟不上地产业的迅速发展，因此，崭新的城市综合体开发模式已成为地产商们的新型课题。

城市综合体是指将城市中的商业、办公、居住、旅店、展览、餐饮、会议、文娱和交通等生活功能融为一体，并使各部分间相互依存、相互受益，从而形成的一种多功能、高效率的建筑群落。由于城市综合体基本具备了现代城市的全部功能，因此又被称为"城中之城"。

成功的城市综合体的发展有几点必须涵盖：首先是独具的创意，符合时代审美观的外在形态；其次，好的规划设计，包括科学合理地安排住宅、酒店式公寓、写字楼，也涵盖商场、酒店、主题公园等多种项目，而这些都要根据当地城市发展进程、家庭收入情况、居民的购买力和消费结构来决定；然后，要做好人流组织、交通组织、空间组合等，不同业态之间要做到互相不扰，使综合体真正发挥紧凑、综合的功能；最后，注重建筑与生态相结合，增加绿化面积，实现人文生态，使综合体中的各种人群和谐相处。

这种城市地标性建筑，正逐渐成为各大城市商业中心区和国际化标准的生活模板，是高品质城市生活的标志；它提升着城市的价值、品质和整体形象，是拉动区域经济和文化的引擎。它的多重复合性决定了其具有强而多的社会功能——优秀的商业综合体能够增强城市的流通力，提高居住人群的生活和消费品质，也能成为地方重要的旅游景点，对于增强城市吸引力有重要作用。其间最主要的一点是立体交通的配合。城市综合体的规划必须与交通等规划设计相协调，交通承载着其发展的重要疏通作用。打造完善的城市公共空间，必然需要稳定流畅的交通枢纽和精神丰富的文化广场等。交通与城市、区域形成联动发展，实现综合体一体化。

现代城市的发展正由粗放型向集约型方向转化，城市综合体越来越多地强调开放性与融合性。日趋完备的商业办公、衣食、交通和文化氛围，促进着多功能区的融合。城市综合体之间相互促进，形成群体城市发展的大好形式，整体城市覆盖率的大幅度提升指日可待。

RAFFLES CITY BEIJING
北京来福士中心

Architects: SUNLAY Architectural Design Ltd.
Chief Design Inspector: Zhang Hua
Location: Beijing, China
Area: 145,928 m²

设计机构：三磊建筑设计有限公司
设计总监：张华
地点：中国北京
面积：145 928 平方米

The diagonal place of this location of Raffles City Beijing is the "Dongzhimen Transportation Hub" which is under construction with superior geographical position and the traffic condition here is very convenient. This location is also a very important commercial plot of the city centre and one of the advanced projects under construction along the East Second Ring Road.

The starting points in designing this project are to provide a high-quality urban plan; perfectly fuse the project development and city transportation system; improve the city quality of this area, thus transforming this location to a new landmark in the city intersection point; fit the market demand and deeply grasp the active adaptability of architects towards industry development; create a vivid visual experience for people; effectively organize the complex function; increase the building utilization rate and finally reap more economic benefits for proprietors.

The design target of the overall layout is to echo with the city planning, create a good architectural landscape and establish a clear functional zone and non-hindrance inner connection. In order to avoid the depressive feeling produced by high-rise architecture blocks (especially for Dongzhimen overpass), the whole building is segmented according to the function defined by the project plan, which makes an architectural block with proper location and proportion.

This project functions consist of several parts, including commerce, office buildings, serviced apartments, club houses, an underground parking lot and supporting houses. The design targets include efficient building performance, more functional choice, multi-technology implementation, space openness and flexibility, and traffic accessibility.

The design target of architectural image is to create harmonious and changeable building image with visual impact and perfectly mix with the city environment.

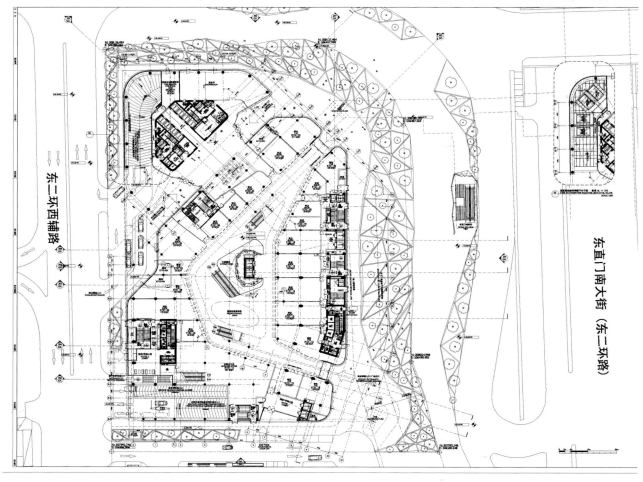

东二环西辅路

东直门南大街（东二环路）

Ground Floor Plan 首层平面图

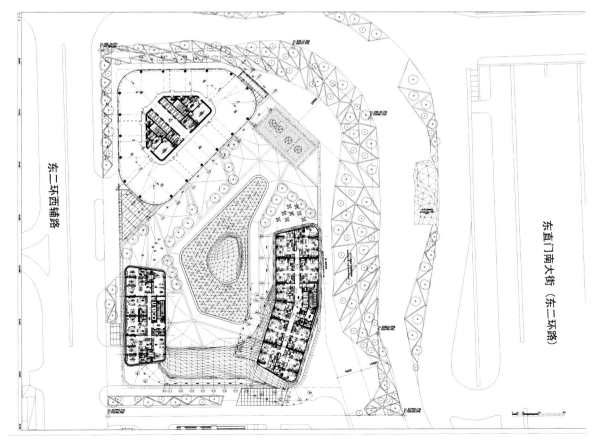

北京来福士广场地理位置优越、交通便捷，是中心区重要的商业地块，也是东二环沿线开发在建项目的高端项目之一。

本项目设计的出发点如下：优质的城市规划为先导；项目开发与城市交通系统发展完美地融合；提升项目所在区域的城市品质，成为城市交叉点上的新地标建筑；适应市场需求，深度把握建筑对产业发展动态的适应性；为人们创造一种鲜活生动的视觉体验；有效地组织复杂的使用功能，提高建筑使用率，为业主创造更大的经济效益等。

本项目的总体布局以与城市规划相呼应、创造良好的城市建筑景观、创建清晰明了的功能分区和无阻的内部联系为设计目标。为了避免过大的高层建筑体量对城市产生压抑感（特别是对东直门立交桥地段），设计师根据项目设计任务书的功能要求，对整体建筑进行了分割，使其成为一组位置合理、比例恰当的系列建筑体块。

该项目功能由商业、办公楼、酒店式公寓、会所、地下车库及配套用房几大部分组成。设计均以实现高效的建筑性能、更多的功能选择、多种科技的普遍利用、空间的开放性及灵活性和交通的可达性为目标。

建筑形象设计致力于打造和谐、变化、富有视觉冲击力的整体建筑形象，并使之能与城市环境完美融合。

General Arrangement Plan Typical Upper Level　标准层平面组合图

ARGANZUELA FOOTBRIDGE
ARGANZUELA 步行桥

Architects: Dominique Perrault Architecture
Client: Madrid City Council
Engineering: MC2 / Julio Martínez Calzón (stucture);
TYPSA (mechanical engineering)
Location: Parque de la Arganzuela,Madrid, Spain
Site Area: 100,000 m²
Building Area: footbridge 150 m (section 1) 128 m (section 2) length,
5 to 12 m width
Photographer: DPA/ADAGP

设计机构：多米尼克·佩罗建筑师事务所
客户：Madrid City Council
工程机构：MC2 / Julio Martínez Calzón（结构工程）;
TYPSA（机械工程）
项目地址：西班牙马德里 Arganzuela 公园
占地面积：100 000 平方米
建设区域：步行桥第一部分长 150,
第二部分长 128 米，5~12 米宽
摄影：DPA/ADAGP

The burying underground of the highway that ran along the edges of the Manzanares river provided the opportunity to open up a new urban territory to the inhabitants of Madrid: the Manzanares Park. A series of bridges over the river will allow passage from one side of the park to the other. Designed to link the neighborhoods on the right and left banks of the river, the Arganzuela Footbridge will be the longest of all the bridges to be built. The bridge will be for both pedestrians and cyclists. The footbridge enables people to cross from one side of the park to the other while also providing direct access to the park below. Cone like in structure, the bridge has two interlocking metal spirals, wrapped by a metallic ribbon. Spaced wooden slats make up the floor of the bridge, allowing the rays of the sun to penetrate through to the park below. The cones' geographic location creates a belvedere over the park and the surrounding city as well as an exceptional location from which it will be possible to admire the famous Toledo Bridge. Shaded during the day, the promenade becomes luminous at night.

曼萨纳雷斯河边的高速公路铺设于地下，这给马德里的居民提供了更加开放的城市新领地，即曼萨纳雷斯公园。一系列横跨在河上的大桥能使人们从公园的这一端走到另一端。Arganzuela 步行桥可以连接曼萨纳雷斯河左岸与右岸的地区，是将建桥梁里边最长的。这座桥为曼萨纳雷斯河两岸提供了一个直接的通道，同时也为逛公园的行人和自行车提供了去往下游区域的路径。人们通过步行桥可以从公园的一侧到另一侧，桥上还有到桥下公园的直接入口。这座桥由一对圆锥结构组成，外部包裹着互相交织的金属丝带。桥面由木板组成，阳光可以穿透桥面直达桥下的公园。锥形的地理位置在公园和周围的城市上方创造了一个观景台，在其上可以欣赏到著名的托莱多桥。步行桥在白天可以遮阳，在夜晚则灯火通明。

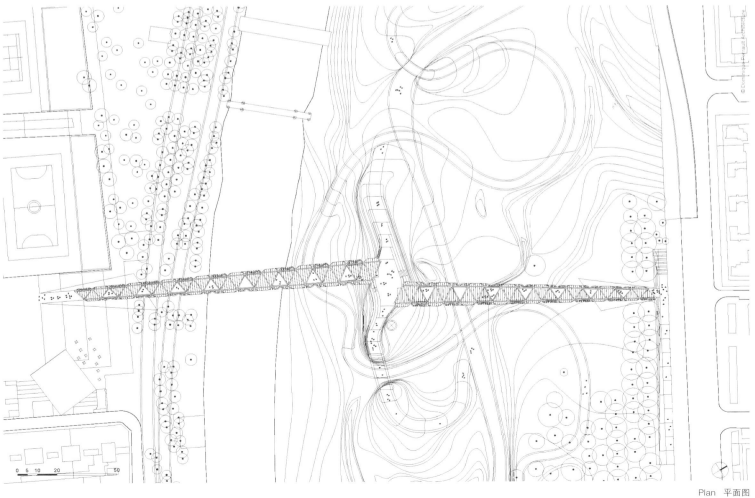

Plan 平面图

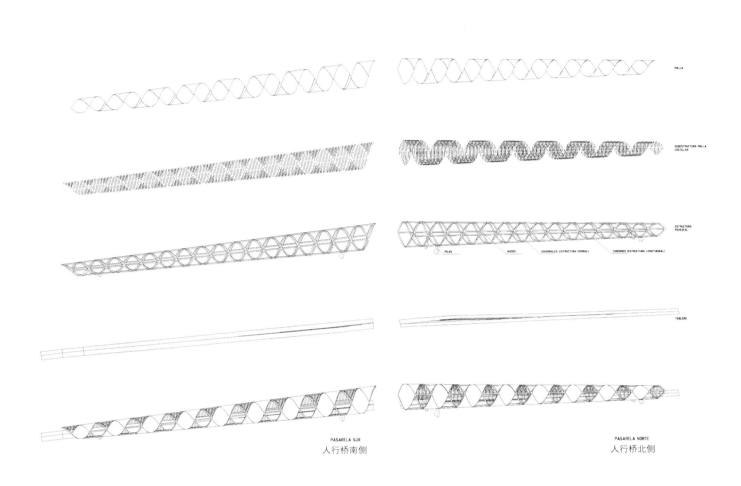

MALLA

SUBESTRUCTURA MALLA
COSTILLAS

ESTRUCTURA
PRINCIPAL

PILAS   NUDOS   DIAGONALES (ESTRUCTURA ESPIRAL)   CORDONES (ESTRUCTURA LONGITUDINAL)

TABLERO

PASARELA SUR
人行桥南侧

PASARELA NORTE
人行桥北侧

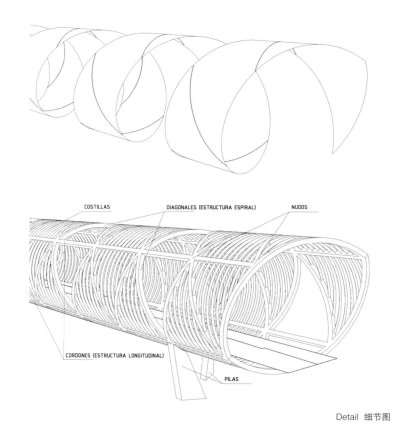

COSTILLAS　　DIAGONALES (ESTRUCTURA ESPIRAL)　　NUDOS

CORDONES (ESTRUCTURA LONGITUDINAL)

PILAS

Detail 细节图

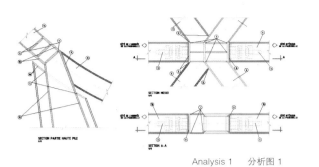

SCHÉMA DU NŒUD TYPE DE LA PILE (NŒUD ARTICULÉ)

Analysis 1　分析图 1

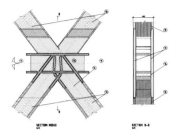

DÉTAIL DU NŒUD ENCASTRÉ

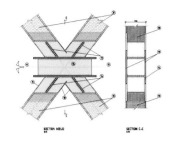

DÉTAIL DU NŒUD DEMI-ARTICULÉ

Analysis 2　分析图 2

TIANJIN WEST
RAILWAY STATION
天津西站

Architects: GMP · von Gerkan, Marg and Partners Architects
Designer: Meinhard von Gerkan and Stephan Schütz with Stephan Rewolle
Client: Tianjin Ministry of Railway
Project Leader: Linlin Jiang
Location: Tianjing, China
Building Area: 179,000 m²
Photographer: Christian Gahl

设计机构：GMP 建筑师事务所
设计师：Meinhard von Gerkan and Stephan Schütz with Stephan Rewolle
客户：天津铁路局
项目经理：蒋琳琳
项目地址：中国天津
建筑面积：179 000 平方米
摄影：Christian Gahl

After a construction period of two and a half years, von Gerkan, Marg and Partners Architects (GMP) have completed the Tianjin West Railway Station in China. The new intersection, which is located about 130 kilometers south-west of Beijing, serves as a stop on the high-speed line between the Chinese capital and Shanghai, as well as connecting the various regional lines and linking these to the underground network. The local urban design function of the railway station is to connect a commercial area to the north with the old city center to the south, bridging tracks, a river and a road in this city of 12 million residents.

The architects have highlighted the bridge function between the city quarters with a 57-metre high and nearly 400-metre long barrel vault roof above the terminal concourse. Its curved roof is reminiscent of a large scale city gate and the long, stretched out concourse beneath of a classic place of transit. The portals of the eastern and western sides of the curved hall are symmetrically framed by arcades. To the south of the building a large and open station forecourt covers a wide area which gives credence to the importance and dimension of this railway station.

Passengers enter the new Tianjin West Railway Station through the main entrances on the north and south sides. Arched cantilevers above the entrances and tall window fronts convey an initial impression of the space passengers encounter in the concourse, which is flooded with daylight, providing a highquality atmosphere and clear orientation for travellers.

Daylight reaches the concourse through the diamondshaped steel and glass roof construction, and while the lower part is nearly transparent and admits a great deal of light, the upper part serves as protection against direct solar radiation. The barrel vault roof conveys a dynamic impression, not least because its steel elements vary in width and depth from the bottom to the top, and they are woven together. Escalators and lifts are available for passengers and visitors to descend to the platforms. This technically and structurally sustainable railway station illustrates a contemporary interpretation of the cathedrals of traffic from the heydays of railway travel.

经过两年半的建设，由 GMP 设计的天津西站终于完工，这里距离北京西南 130 千米，是北京与上海高速铁路以及连接其他线路高速列车的地下站点。西站在当地城市的功能是在这个有一千两百万人口的城市中将北面的商业区和南面的老城区、桥接轨道、河流和道路连接起来。

建筑师在候车室上方设计了一个长 57 米、高 400 米长的穹顶，曲线的屋顶让人联想到大型的城门和地下延伸的运输线。大厅东西两侧的入口呈对称状。建筑的南边有一个开放的大型站前广场，强调了火车站的重要性。

乘客主要通过南北两个入口进出新天津西站。入口上方的拱形悬臂和高大的前窗为大厅里的乘客传达了一种初始的视觉印象，这里充满着阳光，为游客提供了一个高质量的环境和清晰的走向指示。

阳光穿过钻石般的钢铁玻璃屋顶到达大厅，屋顶下层是透明的，可以引入更多光线，上层引光处则做了防辐射处理。穹隆式屋顶充满了动感，这不仅是因为从顶到底的钢铁结构在宽度和高度上充满变化，还因为这些结构交织在一起。自动扶梯和电梯承担乘客上下交通的功能。这种拥有可持续性技术和结构的火车站是现代铁路交通全盛时期的缩影。

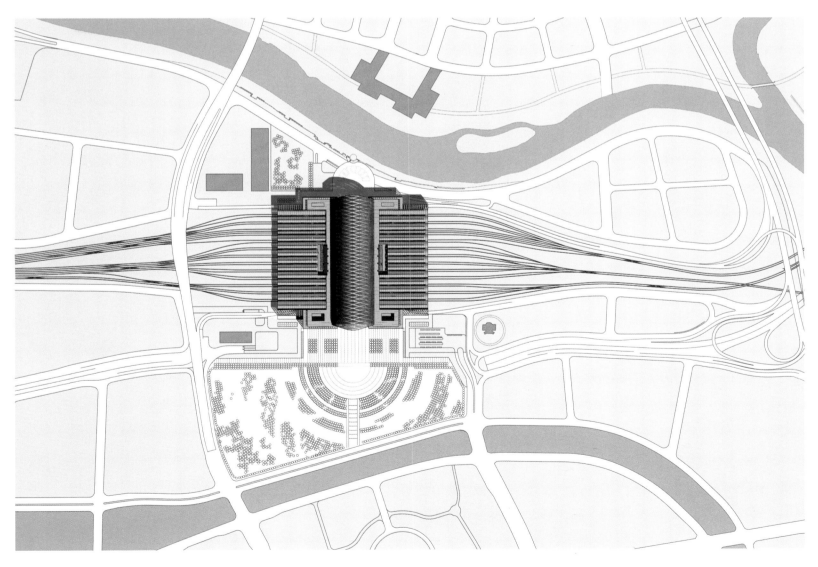

Plan 平面图

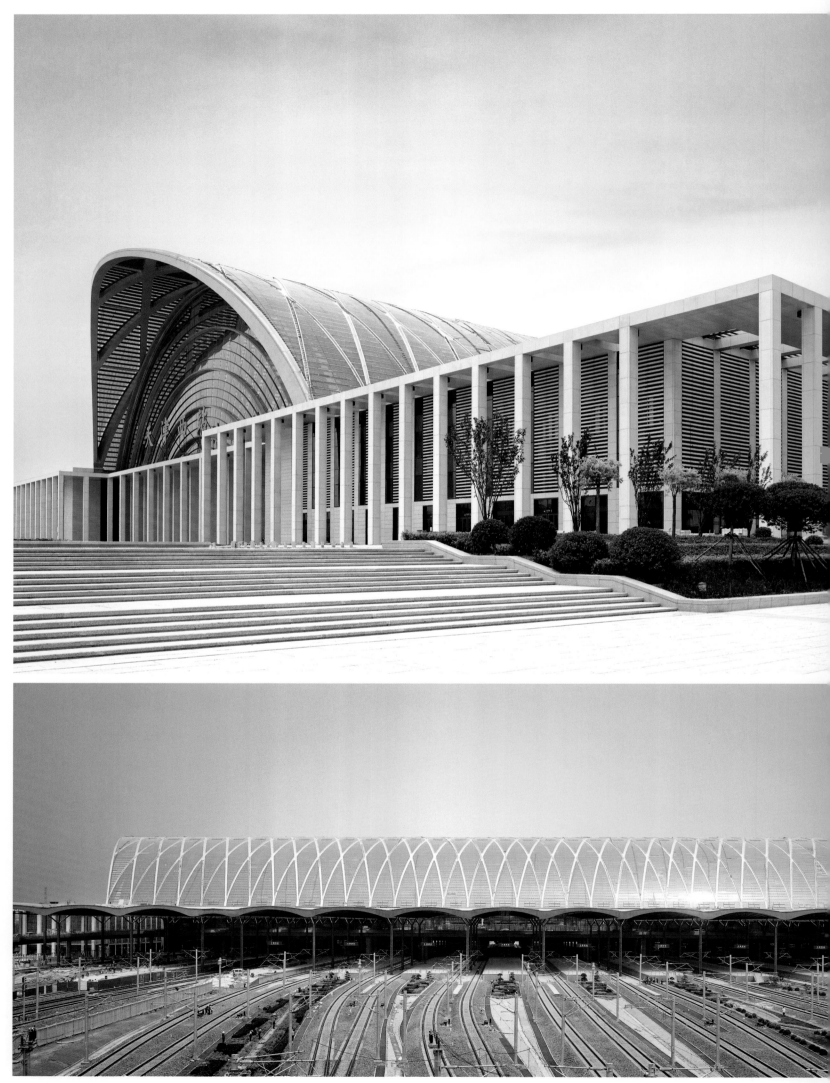

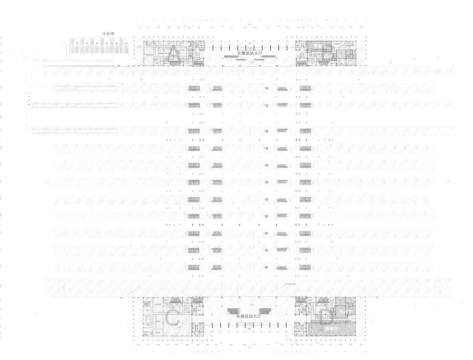

Analysis 1 分析图 1

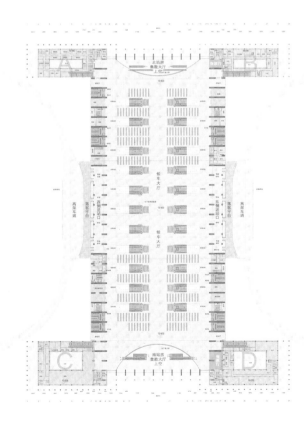

Analysis 2 分析图 2

KING'S CROSS STATION
国王十字火车站

Architects: John McAslan + Partners
Location: London, UK
Area: 7,500 m²
Photographer: Hufton and Crow, Phil Adams

设计机构：John McAslan + Partners
项目地址：英国伦敦
面积：7 500 平方米
摄影：Hufton and Crow，Phil Adams

Western concourse
The Western Concourse sits adjacent to the facade of the Western Range, clearly revealing the restored brickwork and masonry of the original station. From this dramatic interior space, passengers access the platforms either through the ground level gate-lines in the Ticket Hall via the Western Range building, or by using the mezzanine level gate-line, which leads them onto the new cross–platform footbridge.
Located above the new London Underground northern ticketing hall, and with retail elements at mezzanine level, the concourse will transform passenger facilities, whilst also enhancing links to the London Underground, and bus, taxi and train connections at St Pancras. The concourse is set to become an architectural gateway to the King's Cross Central mixed-use developments, a key approach to the eastern entrance of St Pancras International. It will also act as an extension to King's Cross Square, a new plaza that will be formed between the station's southern facade and Euston Road.

Eastern Range
Main Train Shed Roof
Platforms, Footbridge and Subway
Shared Service Yard Plant Room
New Platform
Western Range and Concourse
King's Cross Square

Western range

The Western Range at King's Cross is the historic station's biggest component, accommodating a wide range of uses.Complex in plan, and articulated in five buildings, the practice's considered architectural intervention has delivered greatly improved working conditions for the station staff, train-operating companies and Network Rail management teams. The Northern Wing, destroyed by bombing in World War II, has been rebuilt to its original design.

Main train shed

The station's Main Train Shed is 250m long, 22m high and 65m wide, spanning eight platforms. The restoration includes revealing the bold architecture of the original south facade, re-glazing the north and south gables and refurbishing platforms. The two barrel-vaulted roofs are currently being refurbished and lined with energy-saving photovoltaic arrays along the linear roof lanterns. A new glass footbridge designed by John McAslan + Partners extends across the Main Train Shed, replacing the old mid-shed Handyside bridge and giving access to every platform as well as the mezzanine level of the concourse.

John McAslan + Partners' design integrates the main and suburban train sheds for the first time, creating a completely coherent groundplan for passenger movements into and through the station. Improvements to the Suburban Train Shed is located to the north of the Western Concourse and Western Range buildings have enhanced the operation of its three platforms.

The ambitious transformation of the station creates a remarkable dialogue between Cubitt's original station and 21st century architecture — a quantum shift in strategic infrastructure design in the UK. This relationship between old and new creates a modern transport super-hub at King's Cross, whilst revitalising and unveiling one of the great railway monuments of Britain.

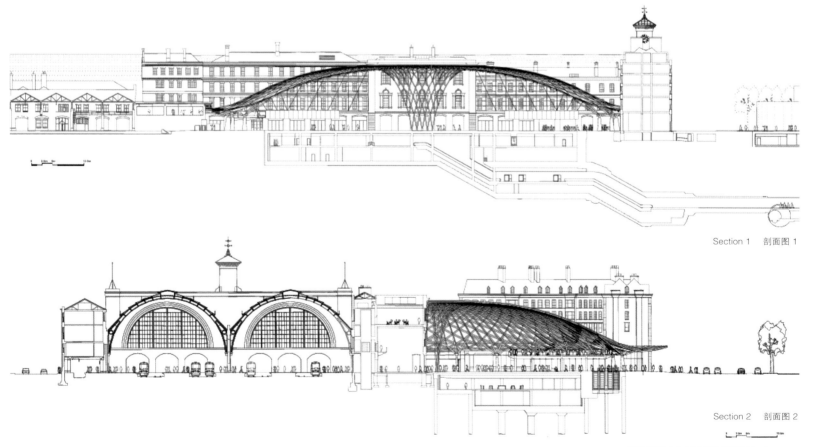

Section 1   剖面图 1

Section 2   剖面图 2

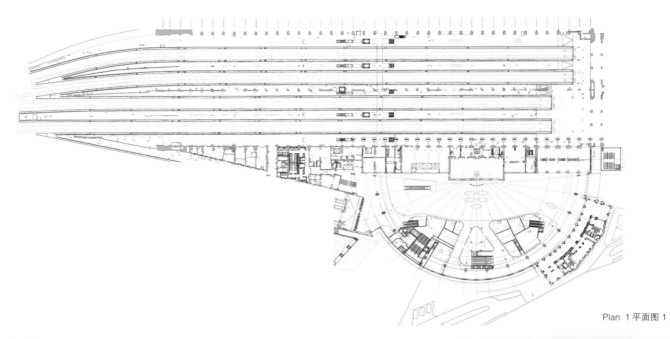

Plan 1 平面图 1

西大厅

西大厅毗邻西环区,清楚地展示着原火车站已修复的砌砖和石块。从这个引人注目的室内空间,乘客可经由西环区穿过售票大厅地面层的入口到达站台,或通过使用中间层的入口直达新的十字路口——站台人行天桥。

西大厅位于新伦敦地铁北售票大厅上方,设有零售店的中间层,为乘客提供了诸多的便利,同时加强了伦敦地铁与圣潘克拉斯公交车、出租车和火车之间的联系。西大厅将成为国王十字中心综合开发区的重要通道以及圣潘克拉斯国际火车站东部主要的入口。它还将成为国王十字广场的扩展腹地,在车站的南部和尤斯顿路之间形成一个新广场。

西环区

国王十字火车站的西环区域使用范围非常广,是这个具有历史意义的火车站的最大组成部分。它的规划很复杂,不仅连接在五栋建筑物之中,同时也为车站工作人员、列车操作公司和铁路网络管理团队极大地改善了工作条件。在第二次世界大战中被轰炸摧毁的北区已经恢复了其原始的设计模样。

主火车棚

主火车棚长250米、高22米、宽65米,横跨8个站台。重建包括恢复原始南侧外立面的粗体结构,为北部和南部的墙壁装配玻璃并翻新修整站台。两个桶状的拱形屋顶目前正在重建,设置了节能的光电阵列,并沿线形屋顶灯罩分布。同时由John McAslan 和 Partners 设计的新玻璃幕墙人行天桥延伸穿过主火车棚,替换了旧有的中棚覆盖的汉迪赛德人行天桥,由此可以通往每个站台以及西大厅的中间层。

John McAslan 和 Partners 的设计第一次将市区和郊区火车棚融合在一起,为乘客出入火车站创建了一个完整连贯的总平面标志图。郊区火车棚的改建地点位于西大厅北部,西环区的建设促进了其中三个站台的运作。

车站华丽的转型在以前的原始火车站和21世纪建筑之间建立起了非同寻常的对话——这是英国战略性基础设施设计中的量子转变。这种新旧关系创建了一个现代化的超级交通枢纽——国王十字火车站,同时将振兴英国的交通,成为其铁路史上的里程碑。

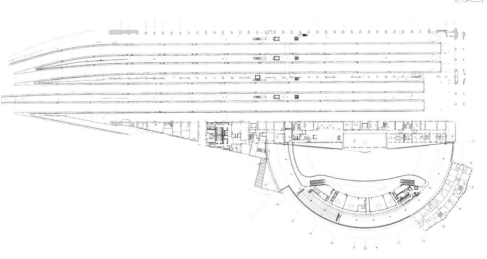

Plan 2 平面图 2

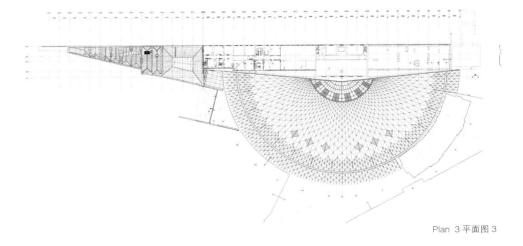

Plan 3 平面图 3

BEIJING SOUTH RAILWAY STATION
北京南火车站

Architects: TPF Farrells
Client : Ministry of Railways (MOR), PRC
Location: Beijing, China
Site Area: 310 000 m^2

设计机构：TPF Farrells
客户：中华人民共和国铁道部
地址：中国北京
占地面积：310 000 平方米

The Beijing South Railway Station is not only a key rail link for China's new high-speed intercity network but it is also a major urban building and master plan. Situated on a 31-hectare site the Station creates an urban link with the surrounding cityscape and acts as a "Gateway to the City" whilst the station itself is designed for a passenger turnover of 105 million passengers annually by the year 2030, with a peak hour flow of 33,280 passengers per hour, and a total of 286,500 passenger movements per day. This integrated design encompasses a multi-modal transport interchange facility with a vertical separation strategy designed so that the passenger traffic flows are direct, convenient and highly efficient.

Completed in 2008 the new station is one of four major rail stations for the new high-speed rail implemented within China and was a core Olympic project endorsed by the Beijing Government and a critical part of Beijing South Railway Station upgrading and extension project.

Situated half a kilometre from the city's old station in Fengtai district between the second and third ring roads, the new Beijing Station will serve as a high-speed intercity rail link which connects Beijing with the Yangtze River Delta cities of Tianjin and Shanghai, with a catchment area of 270 million people.

As the station is immense in scale, the architectural form and structure are clear, simple and people oriented and take into consideration the different operational and management of the various rail lines, station entrances, exits, waiting areas and interchange zones taking place within – the station takes a simple ellipse form that accommodates 3 principle floor levels with two mezzanine floor levels for car-parking and two ancillary gateway office buildings. With such large volumes of passengers it is essential to separate the incoming and departing passengers. One of the main design objectives was to have the passengers board and alight trains with the shortest distance and time possible.

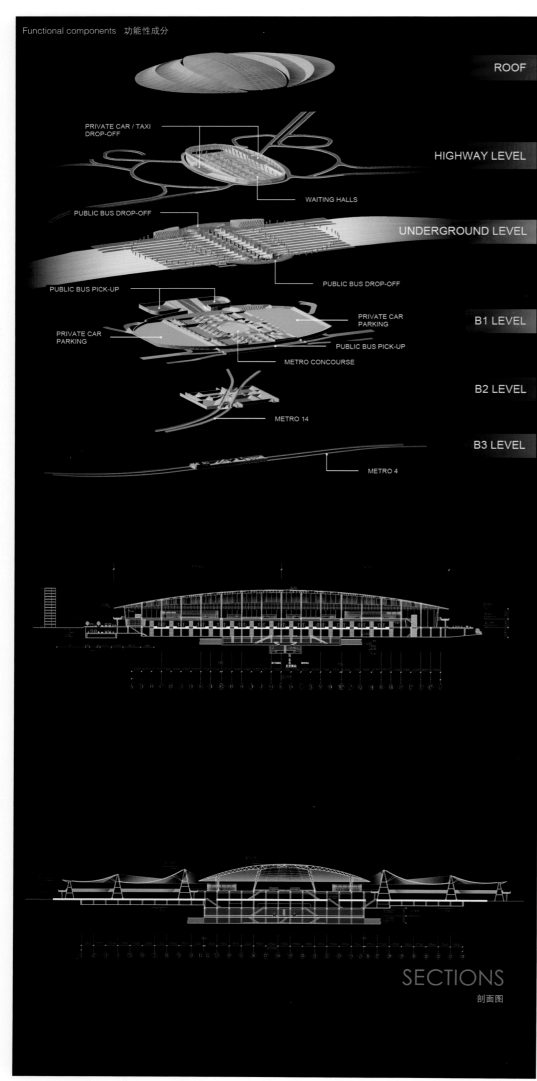

Functional components 功能性成分

ROOF

PRIVATE CAR / TAXI
DROP-OFF

HIGHWAY LEVEL

WAITING HALLS

PUBLIC BUS DROP-OFF

UNDERGROUND LEVEL

PUBLIC BUS DROP-OFF

PUBLIC BUS PICK-UP

PRIVATE CAR
PARKING

B1 LEVEL

PRIVATE CAR
PARKING

PUBLIC BUS PICK-UP

METRO CONCOURSE

B2 LEVEL

METRO 14

B3 LEVEL

METRO 4

SECTIONS

剖面图

The design strategy also incorporates separate zones catering for seamless integration and transition to different types of vehicular traffic including 909 underground basement car-parking spaces, 28 taxi drop-off bays, 24 taxi pick-up bays with 138 queuing spaces and 38 bus spaces (12 drop-off spaces and 26 pick-up bays with 48 queuing spaces) as a comprehensive transport hub. The elliptical plan form is effective in providing an innovative solution to the station's vehicular traffic flow. The overhead road network can adjust to the traffic flows to and from the station area in all directions and assist in relieving the congestion of the surrounding urban arterial roads.

There are a total of 11 island platforms and 2 side platforms with 24 platform edges for High-speed trains (450 metres long), Express trains (500 metres long) and Intercity trains (450 metres long); 2 island platforms with 4 platform edges for the Metro trains(120 metres long) in the basement levels.

As the station is located on existing railway land, the elliptical geometry of the site juxtaposes the diagonal fan of the railway tracks to Beijing's cardinal urban grid. The station scheme creates an urban link with the surrounding cityscape and acts as a "Gateway to the City" with the incorporation of an urban response that unites the railway fan and city grid by inserting a landscaped pedestrian spine in the formal north-south axis that maximizes the sense of approach and creates enhanced public amenity spaces with the northern and southern plazas and the wider city context.

北京南站不仅仅是中国新的城际高速网的一个重要的铁路枢纽,而且还是一个重大的城市建筑和规划项目。车站占地310 000平方米,与周边的城市景观浑然一体,担当着北京的"城市门户",其设计旅客吞吐量为每年1.05亿人次,到2030年时将达到每小时33 280人次的高峰小时流量,每天共计输送旅客286 500人次。车站的集成式设计包含多模式交通换乘设施,采用了垂直分离策略,因而旅客交通流更加直接、方便和高效。

2008年竣工的北京南站是中国新的高速铁路网中四个重点车站之一,是原北京南站改造和扩建工程的关键部分,也是北京市政府重点实施的奥林匹克核心工程。

新北京站距离介于二环路与三环路之间的丰台区的城市老站0.5公里,是连接北京与天津以及上海等长江三角洲城市的一个高速城际交通枢纽,其覆盖区域的人口约为2.7亿。

虽然车站的规模巨大,但建筑形式和结构却十分清晰、简洁,体现了以人为本的理念,同时还充分考虑了车站内各种铁路线路、车站出入口、候车区域、换乘区域的不同营运和管理要求:看似简单的椭圆形车站包含了车站的三个主楼层、两个用于停车的夹层以及两幢附属的办公建筑。对于如此大的客流量,把进站客流和出站客流区分开来是至关重要的。在车站的关键设计目标中,让乘客的上、下车距离和时间达到最短就是其中之一。

设计策略还体现了在不同的区域实现各种交通方式之间的无缝连接和转换这一特色:作为一个综合交通枢纽,该车站具有909个地下停车位、28个出租车落客站、24个出租车上客点(138个排队上车的空间)和38个公交车站(包括12个下客站点、26个上客站点和48个排队上车的空间)。椭圆规划形状可以为车站的车辆交通流量提供有效的创新解决方案。高架路网可以调节各个方向出入车站区域的交通流量,并有助于缓解周边城市干道的拥堵情况。

该车站共有11个岛式站台和2个侧向式站台,设有24个供高速列车(450米长)、普速列车(500米长)和城际列车(450米长)使用的站台面,地下2个岛式站台设有4个供地铁车辆(120米长)使用的站台面。

由于车站位于现有铁路用地上,其几何外形与铁路轨道一样,与北京市的主交通路网斜向相交。该车站采用椭圆形的建筑方案,使其与周边的城市景观浑然一体,形成了北京城的"城市门户",并且通过正南北向的景观人行中轴线把斜向铁路线和城市网格线有机地联系了起来,与城市肌理遥相呼应,提高了进出站的方向感、扩展了南北广场以及更广泛的城市环境的公共活动空间。

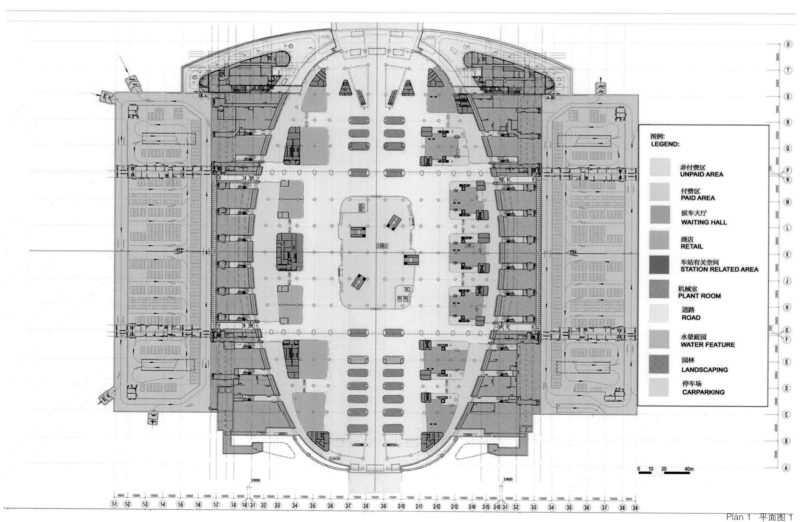

LEGEND section (图例):

图例:
LEGEND:

- 非付费区 UNPAID AREA
- 付费区 PAID AREA
- 侯车大厅 WAITING HALL
- 商店 RETAIL
- 车站有关空间 STATION RELATED AREA
- 机械室 PLANT ROOM
- 道路 ROAD
- 水景庭园 WATER FEATURE
- 园林 LANDSCAPING
- 停车场 CARPARKING

0 10 20   40m

Plan 1 平面图 1

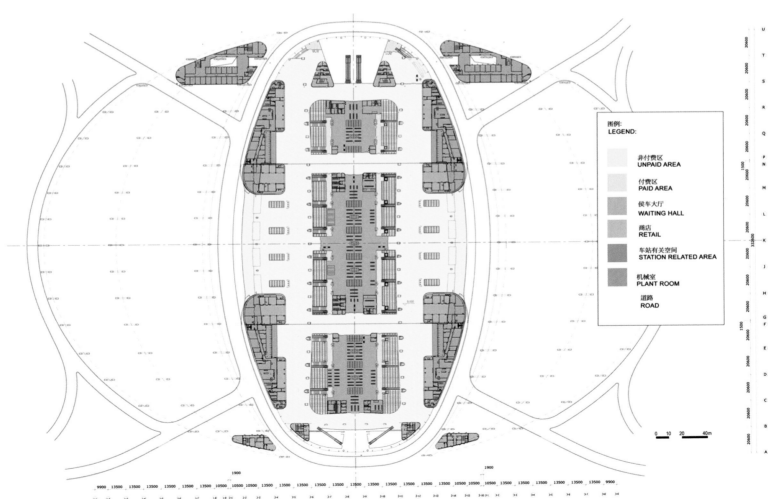

图例:
LEGEND:

非付费区
UNPAID AREA

付费区
PAID AREA

侯车大厅
WAITING HALL

商店
RETAIL

车站有关空间
STATION RELATED AREA

机械室
PLANT ROOM

道路
ROAD

Plan 2　平面图 2

NEW TURIN PORTA SUSA HIGH SPEED TRAIN STATION
新都灵伯塔苏萨高铁站

Architects: AREP
Team: Jean-Marie Duthilleul, EtienneTricaud Silvio d'Ascia architecte,
Project A. Magnaghi architecte
Photographer: Agence S. d'ASCIA

设计机构：AREP
项目团队：Jean-Marie Duthilleul, EtienneTricaud Silvio d'Ascia
architecte, A. Magnaghi architecte
摄影：Agence S. d'ASCIA

Turin Porta Susa High Speed Train Station, the first Italian station on the high speed Paris-Rome line, is resumed at the end of 2009. The gateway into Italy from northern Europe, Turin Porta Susa is designed as an urban locus, an extension of the city's existing Roman layout and public spaces, offering a wide range of transport and services.

The station is an extensive hub where travellers can change easily from one mode of transport — high speed train, regional train, metro, bus, tram, car or two-wheeled vehicle to another. It is also a centre of services and shops for commuters and local residents.

To complete this ambitious project, a mixed-use tower (hotel, offices, public amenities) will be built to the south. It will be accessible to the public and linked directly to the station.

Located between the Spina (the long boulevard crossing Turin from north to south on the site of former railway lines) and the Corso Bolzano, the station is in the form of a long gallery. It is covered by an imposing glass canopy, 385 m long and 30 m wide, which is joined perpendicularly at 100 m intervals by walkways positioned in line with existing streets.

Visitors enter the gallery at these points through large vertical openings in the glass gallery, each protected by a wide canopy. Inside the gallery is a series of volumes in steel and glass, housing services and shops. They rest on a two-level concrete base occupied by car parks and technical areas.

A real street in itself, the covered gallery is bounded to the south by a public amenity tower and to the north by a sloping esplanade linking it to the historic city and the old station (which comprised only a passenger hall). Despite the differences in ground level from north to south and between the Corso Bolzano and the Spina, travellers in the new station can move easily and smoothly between the five levels, by means of a gently sloping ramp, escalators, staircases and lifts.

—Four calvalconi walkways (in blue, at 100-metre intervals) connect the passenger concourse (level - 1) and the platforms below (level - 3).
— Five walkways (in red) allow visitors to cross the gallery from east to west.
Both an intermodal terminal and an urban locus, Torino Porta Susa Station is a highly innovative project in the world of rail transport. It integrates all the requirements of an interchange hub in a new and resolutely contemporary city space: the city enters the station and the station becomes a piece of the city. It can also be seen as a continuation of the urban style of Italy's great 19th-century city arcades (Galleria San Federico in Turin, Galleria Umberto I in Naples, Galleria Vittorio Emanuele II in Milan), and of the great concourses of Europe's 19th-century railway stations.
The skin of the glass roof (15,000 m^2) is entirely covered in single-crystal photovoltaic sensors positioned between the two layers of glass. They also act as shading devices, optimizing the comfort of people using this public space in summer and winter alike. Energy production is 680,000 kWh per year. The whole volume is ventilated naturally — from the platforms, which have high inertia, to the main hall, which is more open to the exterior. The station is temperature controlled in winter and summer.

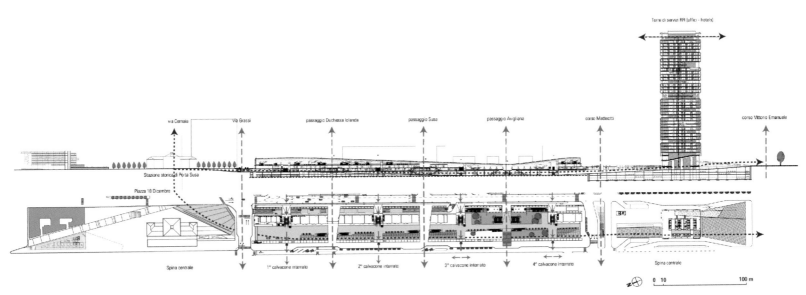

Analysis 分析图

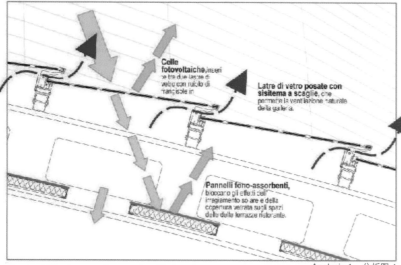

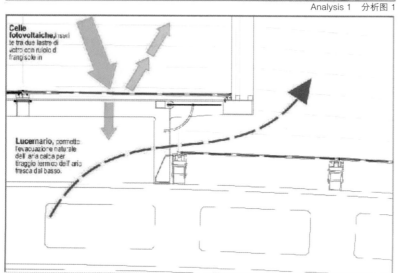

都灵伯塔苏萨火车站是巴黎至罗马的高速铁路线上意大利的第一条高速列车车站，在2009年底重新使用。作为欧洲北部到意大利的门户，都灵伯塔苏萨被规划为城市核心，延伸了现有的罗马式建筑布局和公共空间，提供了范围广泛的交通运输服务。

车站是一个广阔的枢纽，旅客在此可以实现各种交通工具间的方便转换，如高速列车、区域列车、地铁、电车、公共汽车、汽车或两轮车等。它也是一个为当地居民和旅客服务的商业中心。

为完成这项宏伟的计划，车站南边将建设一座多功能大厦（内含酒店、办公室、公共设施），它将向公众开放，并和火车站直接相接。

车站位于Spina（前铁路线旁横穿都灵南北的林荫大道）和博尔扎诺大街之间，像一条长长的走廊。它的上面覆盖有一个长385米、宽30米的玻璃天篷，它通过已有的通道以100米的间隔被连接起来。

游客通过玻璃走廊的垂直入口进入走廊。走廊内部是一系列的不锈钢和玻璃体量，用作服务处和商店。它们位于两层的混凝土基座上，基座则用做停车场和技术区。

这个走廊是一条真正的街道，南边是一个公共设施楼，北边是一个坡地广场，将火车站与这个历史城市和老火车站（现仅存一个旅客大厅）相联系。尽管博尔扎诺大道和Spina之间的南北地面有所不同，但是旅客可以通过坡道、自动扶梯、楼梯和电梯在新火车站里方便地转移，顺利地到达五层中的任何一层。

——四条通道（蓝色，间隔百米）连接负一楼的旅客大厅和负三层的月台。

——五条通道（红色）可使旅客从东到西穿越走廊。

都灵伯塔苏萨车站既是一个多式联运终端又是一个城市核心，在世界铁路运输系统中是一个高度创新的工程。它在一个新的城市空间满足了一个换乘枢纽的所有要求：城市进入车站，车站也成为了城市的一部分。它也可以被视为意大利伟大的19世纪城市商城（都灵圣费德里克商业广场，那不勒斯Umberto商业长廊，米兰Vittorio Emanuele II商业长廊）以及欧洲19世纪火车站恢弘风格的延续。

15 000平方米的玻璃屋顶完全由单晶光电传感器覆盖，位于双层玻璃的中间层。屋顶材料还充当遮阳设备，无论是夏季还是冬季，人们在这个公共空间里都会感到舒适。单晶光电传感器每年产能680 000千瓦每时。整个体量（包括月台、主厅以及更多的外部空间）采用自然通风。车站在冬季和夏季都能对温度进行控制。

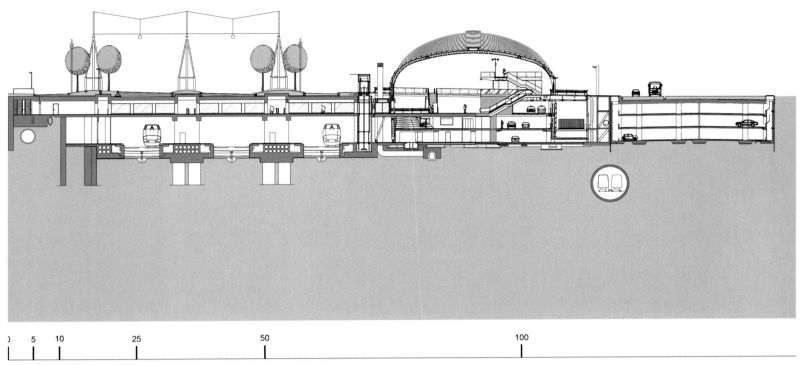

Section 剖面图

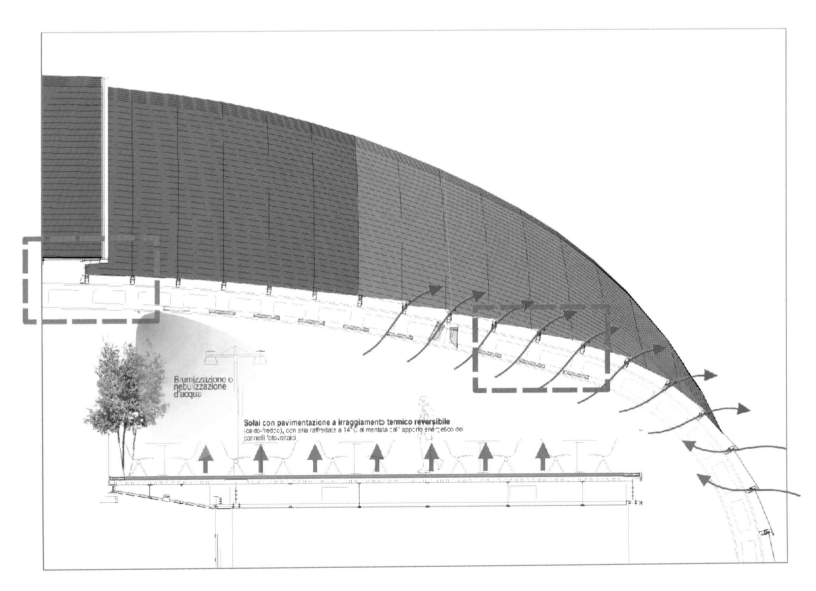

Brumizzazione o
nebulizzazione
d'acqua

Solai con pavimentazione a irraggiamento termico reversibile
(caldo-freddo), con aria raffredata a 14°C al mentata dall'apporto energetico dei
pannelli fotovoltaici

Analysis 3 分析图 3

图书在版编目（CIP）数据

世界建筑事务所精粹：全3册 / 深圳市博远空间文化发展有限公司编 . — 天津：天津大学出版社，2013.5
　　ISBN 978-7-5618-4622-3

　　Ⅰ．①世…　　Ⅱ．①深…　　Ⅲ．①建筑设计—作品集—世界
Ⅳ．① TU206

中国版本图书馆 CIP 数据核字（2013）第 066708 号

世界建筑事务所精粹Ⅰ　　　　深圳市博远空间文化发展有限公司　　编

责任编辑　郝永丽
策划编辑　刘谭春
出版发行　天津大学出版社
出 版 人　杨欢
地　　址　天津市卫津路 92 号天津大学内（邮编：300072）
电　　话　发行部 022-27403647
网　　址　publish.tju.edu.cn
印　　刷　深圳市彩美印刷有限公司
经　　销　全国各地新华书店
开　　本　245 mm×330 mm
印　　张　60
字　　数　810 千
版　　次　2013 年 5 月第 1 版
印　　次　2013 年 5 月第 1 次
定　　价　998.00 元（共 3 册）